W9-ANX-758

WATERLOO HIGH SCHOOL LIBRARY
1464 INDUSTRY RD.
ATWATER, OHIO 44201

The Encyclopedia of US Spacecraft

PRODUCED IN COOPERATION WITH NASA

WATERLOO HIGH SCHOOL LIBRARY
1464 INDUSTRY RD.
ATWATER, OHIO 44201

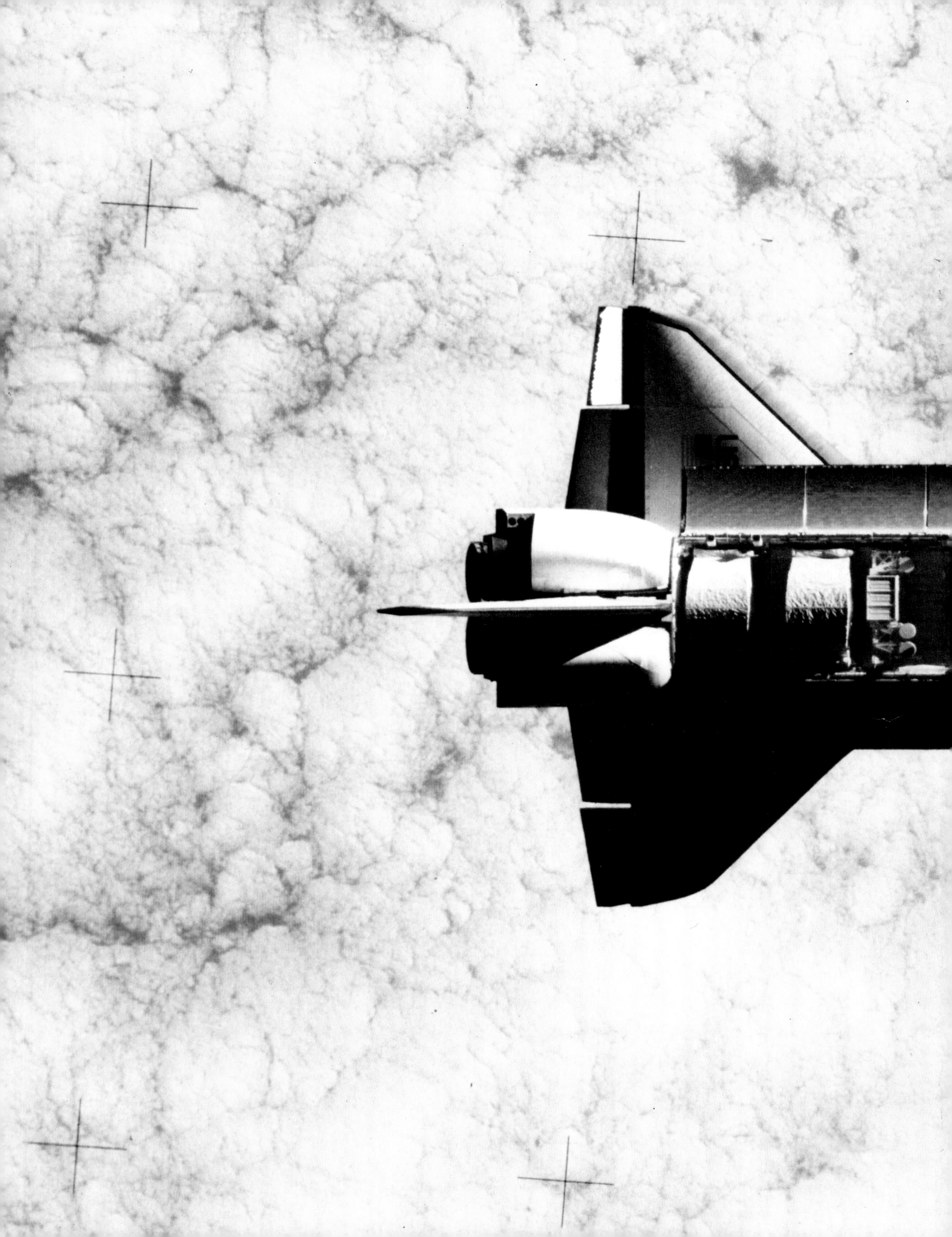

WATERLOO HIGH SCHOOL LIBRARY
1464 INDUSTRY RD.
ATWATER, OHIO 44201

The Encyclopedia of US Spacecraft

PRODUCED IN COOPERATION WITH NASA

Bill Yenne

Exeter Books
NEW YORK
A Bison Book

629.4
YEN

Copyright © 1985 Bison Books Corp.

First published in USA
By Exeter Books
Distributed by Bookthrift
Exeter is a trademark of Simon & Schuster, Inc.
Bookthrift is a registered trademark of Simon & Schuster, Inc.
New York, New York.

ALL RIGHTS RESERVED.

ISBN 0 –671–07580–2.

Printed in Belgium

Picture Credits

All photographs and illustrations were supplied by **the National Aeronautics and Space Administration (NASA),** except for the following:

Boeing Aircraft Company 92-93, 136 (bottom), 158 (bottom left), 179
Hughes Aircraft Company 9 (above right), 30, 34, 36, 57, 67 (left, both), 68, 71, 72 (top), 73 (right), 74 (above left), 76 (both), 77, 78 (both), 89, 102 (top left), 118, 124 (left), 132 (top), 146, 147 (both), 150 (top), 151 (top and bottom right), 158-59, 162-63, 167 (bottom both), 175 (bottom)
Jet Propulsion Laboratory 44, 96 (both), 97, 98-99, 126, 127, 133 (above), 164-65
McDonnell Douglas 58, 60-61, 62-63, 63 (bottom), 103, 105, 106-07, 107 (bottom), 172, 173 (top), 176, 177 (right)
US Air Force 37, 38, 38-39, 39 (right), 90 (bottom), 108, 109 (both), 132 (bottom), 151 (bottom left), 157, 174
© **Bill Yenne** 8 (right), 9 (left and bottom right), 169 (bottom right), 182-83, 190-91

Dedication

The author dedicates this book to Douglas Storer and Gary Deichsel because of their interest and enthusiasm about mankind's great leap into our solar system.

Acknowledgments

The author wishes to thank the following people for helping to make this book possible: Susan Hepner and Susan Flowers at McDonnell Douglas, Sara Kegan at NASA Headquarters, Jack Bateman at Hughes Aircraft and Mary Beth Murrill at the Jet Propulsion Laboratory, with a special tip of the hat to Mike Gentry at the Johnson Space Center and David Thomas at the Goddard Space Flight Center.

Edited by Susan Garratt
Designed by Bill Yenne

Page 1: **Recovery Services for sale! On 14 November 1984 Dale Gardner and Joseph Allen retrieved Westar 6, stranded since its initial deployment. This was seen as a major milestone in the development of the space shuttle's commercial potential.**
Page 2/3: Challenger **over a cloud-covered portion of the earth, photographed during the STS-7 mission on 22 June 1983.**
Below: **James Irwin with the lunar rover, after Apollo 15 touched down on the moon on 30 July 1971.**

Contents

Introduction

Mankind has been fascinated by the heavens for millennia and has entertained notions of space travel for at least a century, but it has only been within the last three decades that it has been possible for man to place his tools, and himself, in the vastness beyond the earth's atmosphere. In the 1980s, when the phrase 'live via satellite' is routine fare on our televisions and space shuttles venture forth with the same regularity as major ocean liners, it is often hard to remember how far we've come since 1957.

In the decade leading up to the launch of the first successful spacecraft, there were still many informed people who considered space flight technically impossible. The successful launch in October 1957 of the Soviet Sputnik spacecraft was a momentous event with global repercussions. The US effort to place a spacecraft in orbit, which had been underway at the time of Sputnik, suddenly became a national priority.

America's first spacecraft, Explorer 1, followed Sputnik by three months, in January 1958, and the race to space was on. Within a year, the United States established the National Aeronautics and Space Administration (NASA) to oversee the development and launch of scientific and communications spacecraft as well as planetary probes. Less than three years after the first spacecraft went into space, both the United States and the Soviet Union put manned spacecraft into outer space. The first manned Mercury spacecraft had barely dried out from its splashdown in the Atlantic when President John F Kennedy outlined America's objectives in space for the next decade during a 21 May 1961 address to Congress:

'Space is open to us now, and our eagerness to share its meaning is not governed by the efforts of others. We go into space because whatever Mankind must undertake, free men must fully share. I believe that this nation should commit itself to achieving the goal, before this decade is out, of landing a man on the moon and returning him safely to the earth. No single space project in this period will be more impressive to mankind or more important to the long-range exploration of space. In a very real sense it will not be one man going to the moon. If we make this judgment affirmatively, it will be an entire nation.'

Above: **An Apollo 16 view of the earth, April 1972.** ***Left:*** **Orbiter** ***Challenger*** **photographed from the unmanned, free-flying German SPAS-01 spacecraft.** ***Below:*** **Apollo 17 CSM during rendezvous and docking maneuvers in lunar orbit.**

The judgment was made affirmatively and the effort to place an American on the moon became the cornerstone of NASA's space program. Orbital flights of the one-man Mercury spacecraft were followed by those of the two-man Gemini, leading up to orbital tests of Apollo spacecraft of the type that would take three men to the moon. Meanwhile, unmanned spacecraft were being sent to the moon to pave the way. The first of these, the Rangers, struck the moon as projectives, transmitting data up to their point of impact. The lunar orbiter program placed spacecraft in lunar orbit, and the Surveyor program soft-landed spacecraft on the moon's surface. All of these spacecraft returned the data about the moon that made the ultimate Apollo landings possible. Kennedy had been dead for six years when the decade ended, but the dream was fulfilled. The United States had landed not one, but two, successful Apollo missions on the moon by the end of 1969.

The last American left the moon in 1972. Following that, NASA's focus on manned space flight was concentrated on the Skylab space station, which was staffed by nine Americans in 1973-74, and the Apollo-Soyuz US/USSR demonstration flight in 1975. For five years thereafter no Americans ventured into space, but in 1981 two flights were made by an American spacecraft that revolutionized space travel. The space shuttle, or Space Transportation System (STS), gave the United States the world's first reusable spacecraft. The space shuttle orbiter can be launched into space and return safely to be relaunched again and again. It can carry seven or more people and its 60-foot cargo bay can be configured to carry a variety of cargo, including other spacecraft that it can release into orbit.

Although manned programs attract the most public attention, the unmanned programs have been far greater in number and of relatively equal importance. These have ranged from weather satellites to communications satellites and have included global photographic survey satellites like Landsat and Seasat. The US has sent spacecraft to orbit the sun and nearby planets. Two Viking spacecraft successfully soft-landed on the planet Mars and returned the first close-up pictures of the Martian surface. Two Pioneer and two Voyager spacecraft have returned spectacular photographs from their flybys of Jupiter and Saturn.

Successful US Lunar and Planetary Landings that returned data from the surface

Year	Landings		
1961			
1962			
1963			
1964			
1965			
1966	Surveyor 1		
1967	Surveyor 3	Surveyor 5	Surveyor 6
1968	Surveyor 7		
1969	Apollo 11	Apollo 12	
1970			
1971	Apollo 14	Apollo 15	
1972	Apollo 16	Apollo 17	
1973			
1974			
1975			
1976	Viking 1	Viking 2	
1977			
1978	Pioneer Venus 2		
1979			
1980			
1981			
1982			
1983			
1984			

Key: Moon, Mars, Venus

Left: **Pioneer 11 spacecraft during a checkout with a mock-up of the third-stage launch vehicle at Cape Kennedy, prior to its launch on 5 April 1973 for the mission to Jupiter. It returned data confirming that Jupiter was mostly hydrogen.**

US Spacecraft

American spacecraft programs fall into three categories: NASA-managed programs, commercial programs and military programs. The NASA-managed programs are primarily scientific in nature. They include spacecraft such as the Explorers, which were used to study the earth's space environment, and probes to other planets. NASA has managed the unmanned landings on the moon, Venus and Mars, as well as all of the US manned space programs, including the Apollo lunar landings. NASA also manages space programs for other civilian government agencies such as the National Oceanic and Atmospheric Administration (NOAA) and the Environmental Science Services Administration (ESSA). In the case of these NASA-managed programs, the spacecraft themselves are sometimes produced by NASA, mainly at the Jet Propulsion Laboratory in Pasadena, California and at the Goddard Space Flight Center near Washington, DC. (Still others are produced by other public institutions such as the Naval Research Laboratory near Washington, DC and the Applied Physics Laboratory at Johns Hopkins University in Baltimore, Maryland.) In other instances, the spacecraft are produced under contract by private manufacturers.

Commercial spacecraft are designed and built by private manufacturers for private users. In some cases the manufacturer and user are the same. For example, RCA builds its own Satcom communications satellites, but Western Union's Westar communications satellites are built under contract by Hughes Aircraft. AT&T built the first ever private spacecraft, Telstar 1, in 1962. When AT&T revived the Telstar designation two decades later, however, it purchased the spacecraft from Hughes. The Hughes HS 376 communications satellite, used by AT&T, is a very good example of a successful commercial spacecraft. Hughes has built 30 and has marketed them to companies in several countries. Nearly all the private-venture spacecraft have been in the area of communications, but with the advent of the versatile space shuttle, private space laboratories and space factories are now planned.

Military spacecraft, like NASA spacecraft, have been built by private customers and developed in-house. In the early days of the space program, there was intense competition between the US military services, but that competition has been replaced with cooperation. The Air Force conducts most major Department of Defense (DOD) launches, including those of Navy spacecraft, and the results of the programs are shared. For example, the Air Force manages the Navstar Global Positioning System (GPS) program but all military forces share its services. In another example, the Air Force launches the Navy's Fleet Satellite Communications System (FLTSATCOM) spacecraft while its own satellite communi-

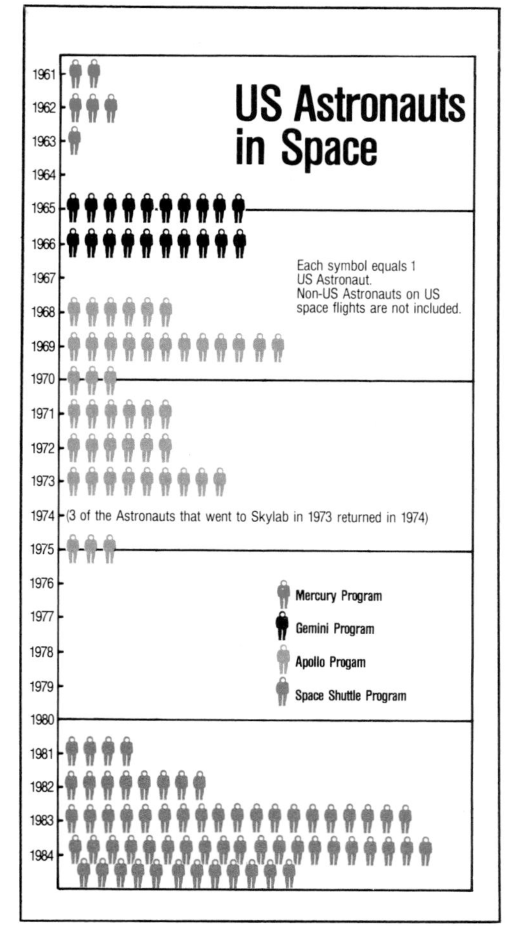

Right: **An Intelsat 4 communications satellite receiving a last look from engineers at the Hughes Aircraft Company. The spacecraft is nearly 18 feet high and almost 8 feet in diameter, and is equipped with spot-beam antennae. The first Intelsat 4 was launched in May 1975.**

cations system (AFSATCOM) program is not a spacecraft at all, but a system attached to the FLTSATCOM.

Launching US Spacecraft

Of the three categories of American spacecraft programs, NASA launches both its own spacecraft and those of commercial clients. The military, usually the Air Force, launches military spacecraft. NASA will occasionally launch a military payload and vice versa, but this is the exception rather than the rule. In the case of the space shuttle program, NASA manages the flights of scientific missions and those where commerical spacecraft are launched or scientific payloads, like the European Space Agency (ESA) spacelab, are carried. If the shuttle mission carries a military payload or is to be used to launch a military satellite, the flight is managed by DOD. Both NASA and DOD use the same types of launch vehicles, and these are discussed in Appendix 2.

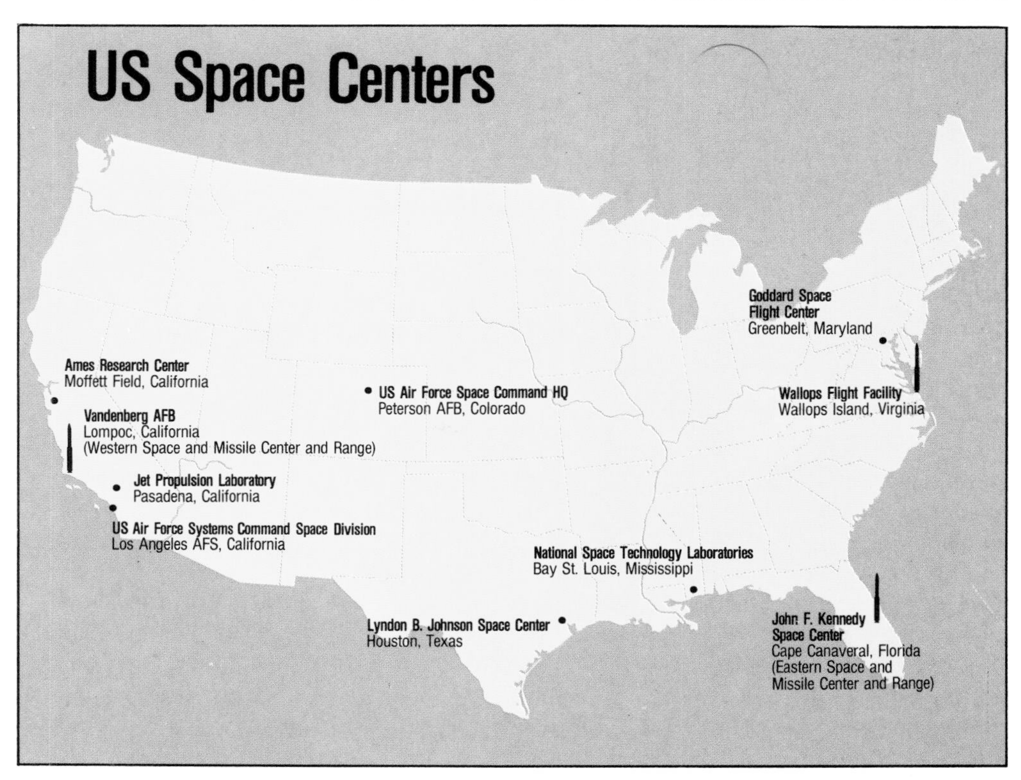

Pad 39A at Kennedy Space Center, Florida has been the launch site for the manned Apollo/Saturn 5 missions to the moon, all the space shuttle and the Skylab space station launches.

American spacecraft are launched primarily from three sites in the United States. These are the Kennedy Space Center at Cape Canaveral, Florida; the Wallops Flight Facility at Wallops Island, Virginia and Vandenberg AFB near Lompoc, California. Beginning in 1970, four Explorer program launches took place at the Italian-operated launch facility near San Marco Island off the coast of Kenya. Some test launches are also conducted from the White Sands Missile Range in New Mexico. Only about 1000 of the over 16,000 US space launches conducted since 1957 have involved orbital missions. Of the spacecraft placed in earth orbit, 42 percent have been launched from the Kennedy Space Center and 55 percent from Vandenberg AFB. Nearly all of the remaining three percent have taken place from Wallops Island, which has been the launch site for 75 percent of US suborbital launches. Most of NASA's launch activity has taken place from Kennedy Space Center, while Vandenberg AFB, a more restricted area, is used for military launches. All of the American planetary probes have been launched from Kennedy Space Center, as have all American manned spacecraft launches. The latter included all the space shuttle launches through 1985, after which a second launch facility was added at Vandenberg AFB for shuttle flights that were managed by DOD.

By the latter half of the 1980s, NASA will gradually lose its monopoly on civilian commercial space launches as the many smaller firms that are now developing launch vehicles put those vehicles into service. Some commercial users, such as General Telephone and Electronics (GTE), have chosen to use the launch services of ESA. ESA launches utilize the French-built Ariane launch vehicle, put aloft from the Guiana Space Center in French Guiana.

The Encyclopedia of US Spacecraft

This encyclopedia is intended to be a detailed catalog of the spacecraft developed in the United States for commercial, military and NASA programs. In preparing this volume, we have concentrated on the physical description of the spacecraft and their systems, together with their launch history. We have also included a brief outline of the spacecraft program and the results obtained in the case of scientific missions.

Commercial and NASA programs have necessarily received more detailed coverage than military programs because many of the latter are classified and only limited data is available. Nevertheless, we have described all of the major military programs, drawing together data from many diverse sources. Data on two closely related topics, space stations and launch vehicles, is included in the appendices.

Re-entry refers to the return of the unmanned spacecraft into the earth's atmosphere at a known time. The re-entry time is noted and no effort is made to recover the vehicle. It either crashes on earth or burns up in the earth's atmosphere.

Recovery takes place when the spacecraft is physically retrieved following its return to the earth's surface.

The Future of US Spacecraft

As the United States enters the final decade and a half of the twentieth century, the nation's first priority in space is the advancement of the commercial use of outer space through the space shuttle programs, together with programs such as shuttle-launched space laboratories and longer-life higher-capacity satellites. The second priority is for a permanent earth-orbiting space station. On the private-venture side, new low-cost launch vehicles might spark the development of a whole new genre of spacecraft. Beyond these immediate goals, other projects are currently under study, such as a permanent (or at least semipermanent) staffed lunar base and further unmanned probes of the outer planets. Also under consideration is a project that has captured the imaginations of scientists and the general public since Americans first ventured into space – a manned mission to Mars.

When considering the future of the US space program beyond the turn of the century, one hesitates to use the phrase, 'the sky's the limit,' because if one thing has been proven by the Americans and their spacecraft, it is that the sky is no limit.

The Syncom IV (Leasat-2) spacecraft moments after departing from the space shuttle *Discovery*'s cargo bay on 31 August 1984.

AEROS

Launch vehicles: Scout
Launch: 16 December 1972 (AEROS A)
16 July 1974 (AEROS B)
Re-entry: 22 August 1973 (AEROS A)
2 September 1975 (AEROS B)
Total weight: 280 pounds (including 62 pounds of payload)
Diameter: 36 inches
Height: 28 inches
Shape: Cylindrical

The AEROS was built with a conical shell at one end, from which a 71-inch probe extended. The flat end was covered with solar cells, which provided power to a nickel-cadmium battery. Power requirements ranged from 4.7 watts to 34.3 watts.

The AEROS launch program was a cooperative effort between NASA and the Bundesministerium für Foschung und Technologie of the Federal Republic of Germany and both were launched from the Western Space and Missile Center at Vandenberg AFB. The mission of AEROS A was to study the state and behavior of the upper atmosphere and ionospheric F region, especially with regard to solar ultraviolet radiation. The mission of AEROS B was to measure the main aeronomic parameters that determine the state of the upper atmosphere and the solar ultraviolet radiation in the wavelength band of main absorption. The orbit achieved varied slightly from that intended, but was deemed acceptable.

A 436-pound AEROS Earth Resources three-axis spin-stabilized satellite is planned by Ball Aerospace/Space America for launch from the space shuttle in 1986.

AFSATCOM

The US Air Force Satellite Communications System (AFSATCOM) has been operational since 19 May 1979 and offers high-priority communications for command and control of US global nuclear forces. The system is a component of the satellites of the US Navy's Fleet Satellite Communications System (FLTSATCOM) and not a system of independent satellites. (See also FLTSATCOM)

Air Density Explorer

(See Explorers 19, 24, 39)

Tests with boiler plate Apollo command modules preceded the manned Apollo CM missions.

Air Density/Injun Explorer

(See Explorer 25)

Air Launched Miniature Vehicle

Launch vehicle: McDonnell Douglas F-15 aircraft
First launch: 24 January 1984 (test)

The Air Launched Miniature Vehicle (ALMV), developed by the Vought Corporation and Boeing Aerospace, was designed to provide the US Air Force with an operational antisatellite (ASAT) capability. More projectile than spacecraft, the ALMV system consists of a modified Boeing short-range attack missile (SRAM) first stage, a Thiokol Altair 3 solid-propellant second stage containing the Vought ALMV with Hughes infrared terminal seeker and conventional high-explosive warhead. When fired, the ALMV will be spun at 1200 rpm to stabilize it with small rocket motors for course correction. The Singer-Kearfott guidance system uses a laser gyro as well as the terminal seeker to locate the target. When operational, the ALMV will be effective against spacecraft with perigees up to 620 miles.
(See also Calsphere, Squanto Terror)

AMPTE

Launch vehicle: Delta 3924
Launch: 16 August 1984
Total weight: 533 pounds (US spacecraft)
1554 pounds (German spacecraft)
169 pounds (British subsatellite)

The Active Magnetosphere Particle Tracer Explorer (AMPTE) program was undertaken in 1972 as a cooperative research effort between the US, Germany and Great Britain. It involved three symbiotic spacecraft launched simultaneously. One major AMPTE experiment was the release of a barium and lithium cloud, which created an artificial comet to examine its interaction with solar wind. This artificial comet was visible from earth off the US Pacific Coast. The German component (developed by the Max Planck Institute under the sponsorship of the German Research and Technology Ministry) is the Ion Release Module, while the US component (developed by Johns Hopkins Applied

Physics Laboratory) is a charge composition explorer that studies the ions released by the German spacecraft. The British subsatellite (developed under the auspices of the British Science and Engineering Research Council) paralleled the course of the others, measuring the effects of the artificial comet on natural space plasma.

AMSAT-OSCAR

Launch vehicle: Carried by ITOS
Total weight: 35 pounds

AMSAT-OSCAR is a series of six amateur-built and -operated satellites carrying an amateur radio and designed to retransmit Morse-code messages. AMSAT was a nonmodular spacecraft powered by nickel-cadmium batteries that provided 24 volts direct current at 3.5 watts. It was 17 inches long, 12 inches wide and 6 inches high.
(See also ITOS, OSCAR)

Apollo

The Apollo program was designed to land a series of American manned expeditions on the surface of the moon and return the astronauts safely to earth. The initial Apollo launches were hardware development tests. Apollo 7 was the first manned Apollo spacecraft and Apollo 11 was the first Apollo spacecraft to effect a lunar landing. All of the manned Apollo launches took place at Kennedy Space Center (KSC).

The Apollo spacecraft was an 82-foot structure launched atop a Saturn launch vehicle. The spacecraft consisted of five distinct parts, the command module (CM), the service module (SM), the lunar landing module (LM), the launch escape system (LES) and the spacecraft-lunar module adapter (SLA). The latter two were jettisoned early in the launch after they had fulfilled their function. The LES acted like an aircraft ejection seat to get the spacecraft safely away from the launch vehicle in case of malfunction on the launch pad or during the early part of the launch. The SLA served as a smooth aerodynamic enclosure for the lunar module during the launch through the atmosphere and to connect the spacecraft to the launch vehicle.

The Apollo spacecraft series was divided into two parts, or blocks. Block 1 was used for ground tests and unmanned launches, while Block 2 included the spacecraft designed to actually fly lunar missions. The two blocks were similar in size and weight, but in Block 1 the LM was a dummy, or 'boiler-plate,' version called the lunar module test article (LTA). It weighed roughly 29,500 pounds compared to the 32,500-pound average weight of the real LM.

The CM, the control center for the spacecraft, provided the living and working quarters for the three-man crew for the entire flight except for the period when two of the crew were traveling in the LM from lunar orbit to lunar surface. The CM consisted of the inner crew compartment (pressure vessel) and an outer heat shield made of heat-dissipating material that fell away during earth entry. The CM was the only part of the spacecraft that returned to earth.

The compact interior of the crew compartment contained the spacecraft control center as well as crew living quarters. The atmosphere of the compartment was 100

Above: **CSM-101 for the first manned Apollo mission.** ***Below:*** **CSM-104 in the vehicle assembly building.**

percent oxygen heated to 70 to 75 degrees Fahrenheit.

The SM, largest of the modules, contained the electrical power subsystem, environmental control subsystem and the propulsion subsystems required for course corrections to put the spacecraft into lunar orbit and to return it to earth. The subsystems were housed in six wedge-shaped systems arranged around the main engine or service propulsion system (SPS), whose propellant consisted of hydrazine plus unsymmetrical dimethylhydrazine as fuel and nitrogen tetroxide as oxidizer. The SM remained connected to the CM, forming the CSM (command plus service modules) unit until just before re-entry to earth, when it was jettisoned.

The LM, or lunar excursion module (LEM), carried two men from the CSM in lunar orbit to the moon's surface. It was not designed aerodynamically like the other modules because it operated only in the vacuum of space and was wrapped in the SLA during launch. The LM was divided into two parts. The first of these, the descent module, was the part that took the astronauts to the surface of the moon. The second was the ascent module, which separated and returned the astronauts to the CSM. The interior was pressurized and air conditioned like that of the CM. The LM served as a lunar base, with living quarters, communications equipment, cameras (including television) and scientific equipment to measure lunar gravity, magnetic field, seismic activity, atmosphere (as expected, there was none) and radiation, and to conduct analyses of lunar soil and rocks. The LM also carried water, oxygen and helium. In later Apollo missions (J series), the LM carried the lunar roving vehicle (LRV) that was used by the astronauts to drive over the surface of the moon.
(See also Lunar Roving Vehicle)

The Apollo spacecraft are listed here chronologically by launch date.

Apollo Transonic Abort
Launch vehicle: Little Joe 2
Launch and recovery: 13 May 1964

The mission was a simulation of the Apollo LES under conditions where high dynamic pressures and transonic speed conditions exist. It was the first launch of the 'boiler-plate' Apollo spacecraft. The spacecraft was launched from the White Sands Missile Range, New Mexico.

Apollo Max Q Abort
Launch vehicle: Little Joe 2
Launch and recovery: 8 December 1964

Above: **CSM-109 is moved to the integrated workstand for mating with the launch adapter.**
Right: **CSM-106 (Apollo 10) is attached to the Saturn 5 launch vehicle via the SLA.**

Continuing the Apollo LES development, this was the first test of the emergency detection system at abort attitude. It was also the first test of the Canard subsystem (for turnaround and stabilization of spacecraft after launch escape) and of the spacecraft protective cover. The spacecraft was launched from the White Sands Missile Range, New Mexico.

Apollo High Altitude Abort
Launch vehicle: Little Joe 2
Launch and recovery: 19 May 1965

The mission was designed for further LES development. The launch vehicle

Apollo: Command/Service Modules

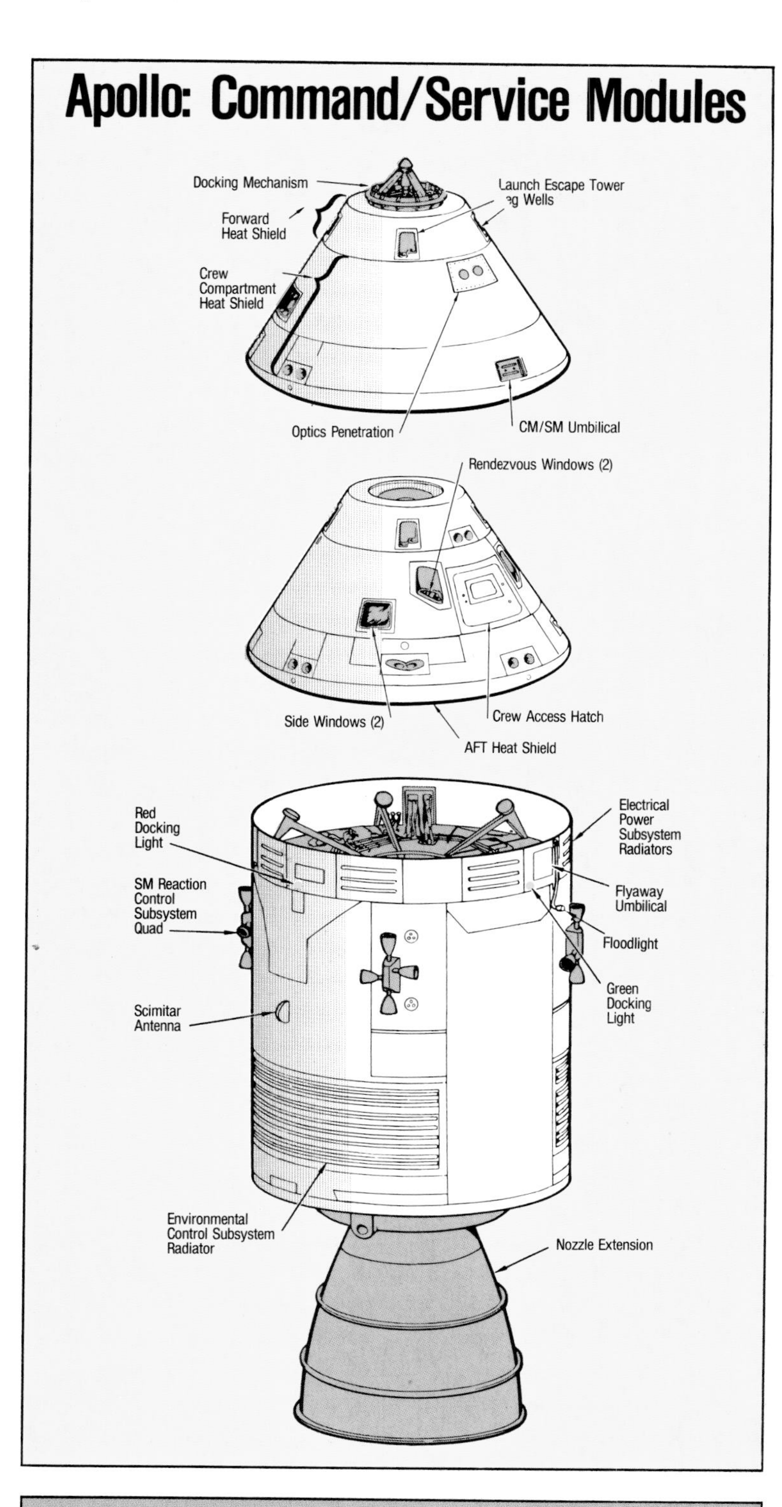

Apollo: Landing Module

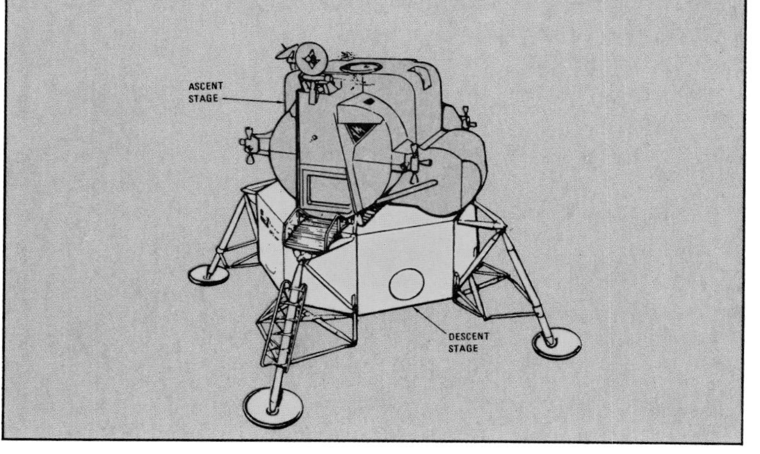

Apollo Spacecraft Command Module (CM)

Height:	12 feet (Apollo 10: 11 feet 5 inches)
Base diameter:	12 feet 10 inches
Shape:	cone
Materials:	Heat shield: brazed stainless-steel honeycomb 0.7-inch thick, filled with phenolic epoxy resin as ablative material. Inner structure: aluminum honeycomb bonded between layers of sheet aluminum 0.25 to 1.25 inches thick.

Apollo Spacecraft Service Module (SM)

Height:	Before Apollo 10: 22 feet Apollo 10: 27 feet 7 inches After Apollo 10: 24 feet 7 inches
Base diameter:	12 feet 10 inches
Shape:	Cylindrical
Materials:	Outer skin: aluminum honeycomb, 1-inch thick Inner structure: milled aluminum radial beams in six sections around a central cylinder

Apollo Spacecraft-Lunar Module Adapter (SLA)

Height:	28 feet
Base diameter:	21 feet
Shape:	Truncated cone
Materials:	Aluminum honeycomb, 1.75-inch thick

(See bottom photo on facing page)

Apollo Spacecraft Lunar Module (LM) Ascent Stage

Height:	12 feet 4 inches (Crew compartment: 3 feet 6 inches)
Base diameter:	14 feet 1 inch (Crew compartment: 7 feet 8 inches)
Shape:	Octagonal (Crew compartment: cylindrical)
Materials:	Thermal and micrometeoroid shield: mylar-aluminum alloy

Apollo Spacecraft Lunar Module (LM) Descent Stage

Height:	10 feet 7 inches
Base diameter:	14 feet 1 inch (31 feet with legs extended)
Shape:	Octagonal
Materials:	Landing gear struts (legs): crushable aluminum honeycomb Thermal and micrometeoroid shield: mylar-aluminum alloy

Above left: **The Apollo 8 CSM was the second manned flight of NASA's lunar landing program.** ***Above:*** **Exterior of 'Apollo 1' after the fire at KSC that killed Grissom, White and Chaffee.** ***Right:*** **Apollo CM-011 is mated to the LM adapter.**

developed a high spin during early powered flight and eventually disintegrated. The LES satisfactorily sensed vehicle malfunction and separated the spacecraft without damage. The high-altitude abort test objectives were not met. The spacecraft was launched from the White Sands Missile Range, New Mexico.

Apollo Saturn (Apollo 1)
Launch vehicle: Saturn 1B
Launch and recovery: 26 February 1966

The first Saturn launch vehicle development flight was unmanned and suborbital. It demonstrated the compatibility and structural integrity of the spacecraft/launch vehicle configuration and evaluated heat-shield performance at a high heating rate. The command module was recovered.

Apollo Saturn (Apollo 2)
Launch vehicle: Saturn 1B
Launch and re-entry: 5 July 1966

A liquid-hydrogen vehicle-development mission, Apollo 2 was to evaluate the fourth stage vent and restart capability and test separation and cryogenic storage at zero G. The flight was terminated during liquid-hydrogen pressure and structural testing.

Apollo Saturn (Apollo 3)
Launch vehicle: Saturn 1B
Launch and recovery: 25 August 1966

The Apollo 3 mission was unmanned and suborbital, providing continued testing of the command/service module subsystems and space vehicle structural integrity and compatibility. The 1-hour 23-minute flight evaluated heat-shield performance at high heat load. Command module 011 was recovered near Wake Island.

Apollo 4
Launch vehicle: Saturn 5
Launch and recovery: 9 November 1967

The 8-hour 30-minute mission was the first Saturn 5 launch and was designed to demonstrate launch vehicle capability and spacecraft development. CSM-017 tested the Apollo heat shield and simulation of the new hatch at lunar re-entry velocity. It was the first launch from Complex 39 at KSC. There were two orbits of 88.3 minutes, after which the vehicle was boosted to a 1068-mile apogee, with re-entry and recovery near Hawaii.

Apollo 5
Launch vehicle: Saturn 1B
Launch: 22 January 1968
Re-entry: 22 January 1968 (LM ascent stage)
12 February 1968 (LM descent stage)

This was the first flight test of the Apollo LM. The flight verified ascent- and descent-stage propulsion systems, including restart and throttle operations. It also evaluated LM staging and fourth-stage orbital performance.

Apollo 6
Launch vehicle: Saturn 5
Launch and recovery: 4 April 1968

The mission for Apollo 6 (CSM-020), the last unmanned Block 1 Apollo flight, was primarily a further launch vehicle testing. Number 2 engine, second stage, malfunctioned and number 3 was shut down by mistake. All three remaining engines were fired at full power, but correct orbit was not achieved. The third stage that was intended to put the spacecraft into lunar orbit failed to ignite, so the spacecraft was separated and the service propulsion system (SPS) was tested. Except for minor problems, all spacecraft systems, including the new crew hatch, performed normally throughout the mission.

The new crew hatch was installed after the disastrous 'Apollo 1' fire of 27 January 1967 at KSC that killed astronauts Virgil Grissom, Edward White and Roger Chaffee in an Apollo capsule.

Apollo 7
Launch vehicle: Saturn 4
Launch: 11 October 1968
Re-entry: 22 October 1968
Crew: Walter Cunningham, Donn Eisele, Walter Schirra

The first manned Apollo flight (CSM-101), this mission had eight successful SPS firings and seven live television sessions. Simulated docking, using the SLA, followed systems checkouts. Two photographic sessions and three medical experiments were conducted. Splashdown in the Pacific came after 260 hours 8 minutes.

Apollo 8
Launch vehicle: Saturn 5
Launch: 21 December 1968
Recovery: 27 December 1968
Crew: William Anders, Frank Borman, James Lovell, Jr

The spacecraft systems were the same as Apollo 7, but this was the first manned Apollo (CSM-103) flight launched by a Saturn 5 vehicle, and the first to achieve lunar orbit (on Christmas Eve). The mission included testing of support facilities, spacecraft navigation communications and midcourse corrections. Splashdown in the Pacific came after 147 hours 11 seconds.

Apollo 9
Launch vehicle: Saturn 5
Launch: 3 March 1969

Left: **The Apollo 9 LM landing gear is deployed, with sensors extending from the landing-gear.**
Above: **Apollo 7 lifts off from KSC, behind the pad service structure and tracking antennae.**
Right: **The historic Apollo 11 launch in 1969.**

Apollo 9 LM upper hatch and docking tunnel.

Recovery: 13 March 1969
Crew: James McDivitt, Russell Schweickart, David Scott

This was the first manned flight of the complete Apollo spacecraft including the LM. The mission was to test the entire integrated system, especially a manned test flight of the independent LM, and to rendezvous between the LM and CSM (CSM-104). Two crew members conducted extravehicular activities (EVAs) and a simulated LM rescue was conducted. The flight was successful and all objectives were met. Splashdown in the Atlantic came after 241 hours 1 minute.

Apollo 10
Launch vehicle: Saturn 5
Launch: 18 May 1969
Recovery: 26 May 1969
Crew: Eugene Cernan, Thomas Stafford, John Young

The mission of Apollo 10 was to conduct all phases of Apollo spacecraft operations except the actual lunar landing, including rendezvous and docking between the CSM and LM in lunar orbit and descent of the LM to within 50,000 feet of the moon's surface. The SPS was restarted on the 31st lunar orbit, bringing the mission to a Pacific splashdown after 192 hours 3 minutes.

Apollo 11
Launch vehicle: Saturn 5
Launch: 16 July 1969
Recovery: 24 July 1969

Apollo Command/Service Modules (CSMs)
Prime Contractor: Rockwell

	Designation (CSM)	Code Name (CSM)	Weight with Fuel (SM)	Weight with Fuel (CM)
Block 1 Spacecraft (unmanned)				
Apollo 4	CSM-017	–	55,000	12,000
Apollo 5	no CSM	no CSM	no CSM	no CSM
Apollo 6	CSM-020	–	42,600	12,500
Block 2 Spacecraft (manned)				
Apollo 7	CSM-101	–	49,730	12,659
Apollo 8	CSM-103	–	51,258	12,392
Apollo 9	CSM-104	*Gumdrop*	36,159	12,405
Apollo 10	CSM-106	*Charlie Brown*	51,371	12,277
Apollo 11	CSM-107	*Columbia*	51,243	12,250
Apollo 12	CSM-108	*Yankee Clipper*	51,105	12,365
Apollo 13	CSM-109	*Odyssey*	51,099	12,327
Apollo 14	CSM-110	*Kitty Hawk*	51,744	12,694
Block 2 Spacecraft: Apollo J (manned)				
Apollo 15	CSM-112	*Endeavor*	54,044	12,774
Apollo 16	CSM-113	*Casper*	54,044	12,874
Apollo 17	CSM-114	*America*	54,044	12,800
Block 2 Spacecraft: Apollo/Skylab missions (manned)				
Apollo/Skylab 2	CSM-116	–	17,000	13,200
Apollo/Skylab 3	CSM-117	–	17,000	13,200
Apollo/Skylab 4	CSM-118	–	17,000	13,200
Block 2 Spacecraft: Apollo-Soyuz Test Project (manned)				
ASTP	CSM-119	*Apollo*	17,676	13,105

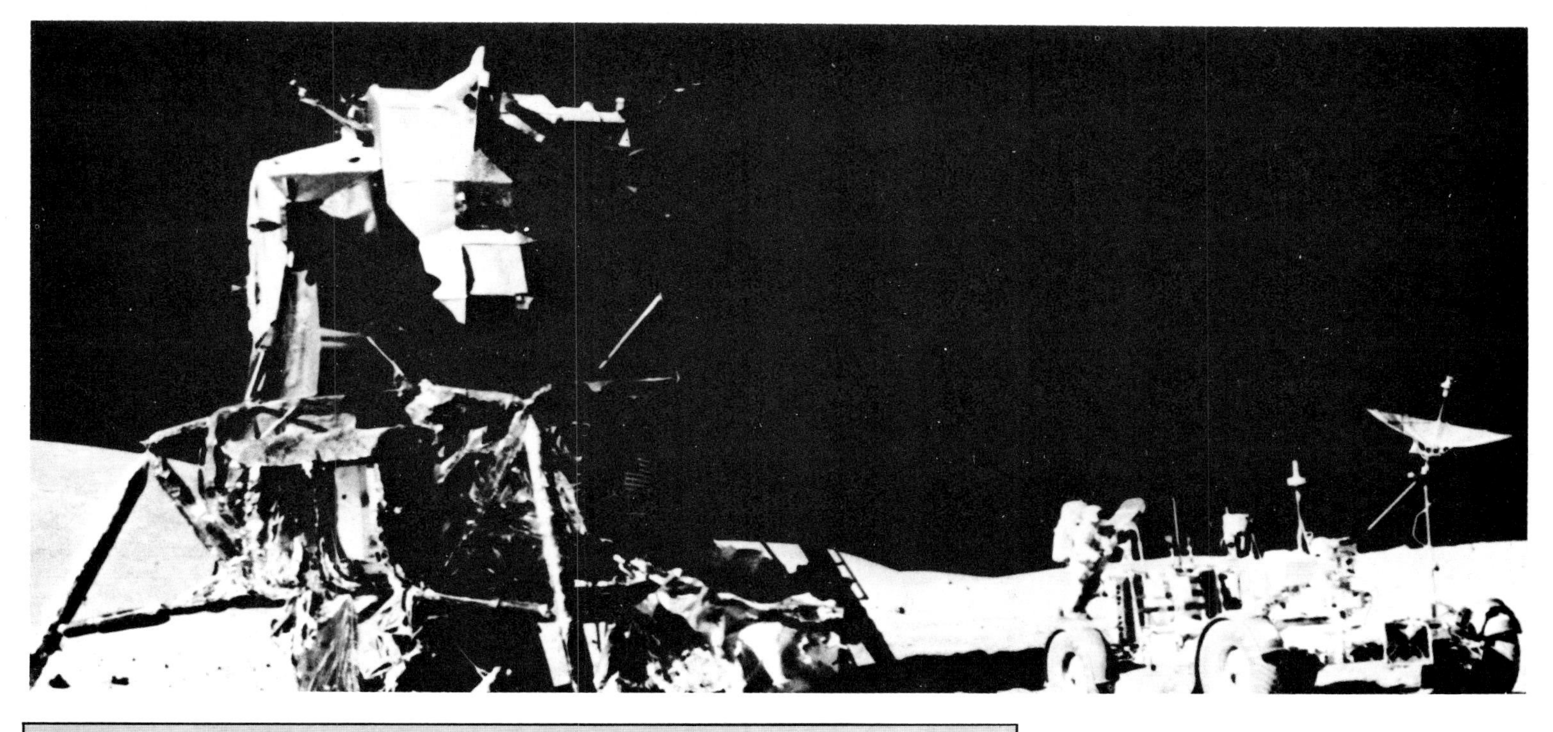

LM-11 Orion with lunar rover on the moon.

Apollo Landing Modules (LM) or Lunar Module Test Article (LTA)
Prime Contractor: Grumman

	Designation	Code Name	Weight with Fuel
Block 1 Spacecraft (unmanned)			
Apollo 4	LTA-1R	–	29,500 (dummy)
Apollo 5	LM-1	–	31,700
Apollo 6	LTA-2R	–	26,000
Block 2 Spacecraft (manned)			
Apollo 7	–	–	none
Apollo 8	LTA-B	–	ballast only
Apollo 9	LM-3	*Spider*	32,021
Apollo 10	LM-4	*Snoopy*	30,849
Apollo 11	LM-5	*Eagle*	33,205
Apollo 12	LM-6	*Intrepid*	33,325
Apollo 13	LM-7	*Aquarius*	32,124
Apollo 14	LM-8	*Antares*	33,680
Block 2 Spacecraft: Apollo J (manned)			
Apollo 15	LM-10	*Falcon*	36,230
Apollo 16	LM-11	*Orion*	36,218
Apollo 17	LM-12	*Challenger*	36,262
Block 2 Spacecraft: Apollo/Skylab missions (manned)			
Apollo/Skylab 2	–	–	none
Apollo/Skylab 3	–	–	none
Apollo/Skylab 4	–	–	none
Block 2 Spacecraft: Apollo-Soyuz Test Project (manned)			
ASTP	–	–	none

Crew: Edwin Aldrin, Neil Armstrong, Michael Collins

This mission conducted the first manned landing on a celestial body other than earth, touching down on the surface of the moon on 20 July 1969. Apollo 11 entered lunar orbit on 19 July, 75 hours 50 minutes into the mission. At 101 hours 12 minutes, Aldrin and Armstrong in LM-5 (*Eagle*) separated from CSM-107 (*Columbia*) to make their descent to the lunar surface, where they landed in the Sea of Tranquility (0° 4′15″ N lat/23° 26′ E long) at 102 hours 45 minutes into the mission. Neil Armstrong became the first human being to set foot on the moon at 109 hours 42 minutes, and was followed by Aldrin 20 minutes later. The total elapsed time spent by the two men outside the spacecraft during EVA was 2 hours 30 minutes, ending at 111 hours 39 minutes. They both walked about 300 feet from the LM, gathered 44 pounds of rock samples and conducted other scientific evaluations. At 124 hours 22 minutes, the *Eagle*'s ascent stage was fired and *Eagle* re-docked with *Columbia* at 128 hours 3 minutes. The LM was separated from the CSM, which returned to earth and splashed down in the Pacific at 195 hours 18 minutes.

Apollo 12
Launch vehicle: Saturn 5
Launch: 14 November 1969
Recovery: 24 November 1969
Crew: Alan Bean, Charles Conrad, Richard Gordon

The mission made a precision landing on the lunar surface in the Sea of Storms on 19 November at 110 hours 32 minutes, 535 feet northwest of the unmanned lunar lander, Surveyor 3, which had soft-landed on 20 April 1967. Surveyor was examined, photographed and its 17-pound television camera and other parts recovered.

The Apollo Lunar Surface Experiments Package (ALSEP) was deployed and left on the moon's surface to gather scientific, seismic and engineering data over a long period. The mission collected 75 pounds of lunar material and ended after 244 hours 36 minutes with splashdown in the Pacific.

Apollo 13
Launch vehicle: Saturn 5
Launch: 11 April 1970
Recovery: 17 April 1970
Crew: Fred Haise, James Lovell, John Swigert

The intended mission was to effect the third soft landing on the moon, but the landing was aborted because of systems failure. At 55 hours 55 minutes into the mission, after a trouble-plagued launch, an electrical short circuit caused an explosion and fire that resulted in failure of the spacecraft's number two oxygen tank and loss of some electrical power. At

Above: **Exterior of the used CSM-110 *Kitty Hawk* (Apollo 14) after splashdown. *Far left, above:* View of earth from Apollo 8. *Far left:* CSM-109 *Odyssey* showing panel damage. *Left:* Apollo 16 LM over the moon prior to docking.**

61 hours 30 minutes, the spacecraft had reached the moon and effected a mid-course maneuver to put it on a return-to-earth trajectory. Although the life support system of the CM was damaged, there was just enough oxygen available in the LM *Aquarius* to use it as a 'lifeboat' to get the crew back to earth safely, where they splashed down in the Pacific.

Apollo 14
Launch vehicle: Saturn 5
Launch: 31 January 1971
Recovery: 9 February 1971
Crew: Edgar Mitchell, Stuart Roosa, Alan Shepard

The Apollo 14 spacecraft (CSM-110 *Kitty Hawk* and LM-8 *Antares*) weighed more than any previous Apollo craft because of safety modifications introduced in the wake of the nearly disastrous Apollo 13 flight. *Antares* landed in the Fra Mauro Crater, 110 miles east of Apollo 12 on 5 February. The astronauts spent a record of 9 hours 24 minutes outside the LM on the surface deploying the ALSEP and collecting 94 pounds of lunar material

WATERLOO HIGH SCHOOL LIBRARY
1464 INDUSTRY RD.
ATWATER, OHIO 44201

for 187 projects in the US and 14 foreign countries. The *Kitty Hawk* returned to earth for a Pacific splashdown 216 hours 2 minutes after launch.

Apollo 15
Launch vehicle: Saturn 5
Launch: 26 July 1971
Recovery: 7 August 1971
Crew: James Irwin, David Scott, Alfred Worden

The fourth successful manned lunar landing, this was also the first of the Apollo J missions capable of a longer stay on the lunar surface and the first to carry the LRV. The total EVA time on the lunar surface was a record 18 hours 37 minutes. Extensive documentation of geologic features was accomplished and 173 pounds of lunar samples were collected. Other activities included deployment of the ALSEP array and drilling for a core sample 10 feet below the lunar surface. The lift-off of the ascent module of the *Falcon* LM-10 was the first seen on earth via television. *Falcon* rejoined the CSM on the latter's 50th lunar orbit and the particles and fields (P&F) subsatellite was released on the 74th orbit. On the return to earth, Worden conducted the first deep-space EVA, collecting film cassettes from the SM. Splashdown was in the Pacific after a record 295 hours.
(See also Apollo P&F Subsatellite, Lunar Roving Vehicle)

Apollo 16
Launch vehicle: Saturn 5
Launch: 16 April 1972
Recovery: 27 April 1972
Crew: Charles Duke, Thomas Mattingly, John Young

The second of the Apollo J series missions included a total of 20 hours 14 minutes of EVA on the lunar surface in which Duke and Young deployed the ALSEP and conducted extensive surface travel aboard the LRV. During the surface exploration new lunar terrain data was obtained, along with 209 pounds of lunar samples. Because of guidance problems and yaw oscillations encountered in the SM propulsion system prior to the landing, NASA decided beforehand to cut the mission short. Having released a P&F satellite in lunar orbit, the CSM returned to earth a day early, splashing down after 265 hours 51 minutes.
(See also Apollo P&F Subsatellite, Lunar Roving Vehicle)

Left: **Equipment transporter tracks lead away from the Apollo 14 LM, *Antares*.**
Right: **Apollo 16 CSM after undocking from the LM.**

Apollo 17
Launch vehicle: Saturn 5
Launch: 7 December 1972
Recovery: 19 December 1972
Crew: Gene Cernan, Ronald Evans, Harrison Schmidt

The third of three Apollo J-series missions was also the sixth and final successful American lunar landing. The ALSEP was deployed, and Cernan and Schmidt conducted an extensive survey of the lunar surface aboard the LRV. Soil samples taken at Shorty Crater revealed for the first time orange-colored material in the soil. The last human activity on the moon's surface during the Apollo program came at 5:23 pm EST on 13 December 1972, with lift-off of the LM ascent module at 5:55 pm. Splashdown in the Pacific occurred after 301 hours 51 minutes.
(See also Lunar Roving Vehicle)

Apollo/Skylab 2
Launch vehicle: Saturn 1B
Launch: 25 May 1973
Recovery: 22 June 1973
Crew: Charles Conrad, Joseph Kerwin, Paul Weitz

Above: **The Apollo 12 LM, *Intrepid*, with the S-band antenna in the foreground.**

Below: **The Apollo 17 LM ascent stage against the blackness of space.**

This was the second Skylab mission and the first use of the manned Apollo spacecraft to take personnel to the Skylab space station that had been launched on 14 May 1973 (Skylab 1). Although the Skylab space station had been damaged in launch and created problems for the crew, the Apollo CSM-116 spacecraft functioned as specified, carrying them to the space station and returning them to earth 28 days later. (See also Appendix 1: Skylab Space Station)

Apollo/Skylab 3
Launch vehicle: Saturn 1B
Launch: 29 July 1973
Recovery: 25 September 1973
Crew: Alan Bean, Owen Garriott, Jack Lousma

This mission was the second use of the Apollo spacecraft to take a crew to the Skylab space station.
(See also Appendix 1: Skylab Space Station)

Apollo/Skylab 4
Launch vehicle: Saturn 1B
Launch: 16 November 1973
Recovery: 8 February 1974
Crew: Gerald Carr, Edward Gibson, William Pogue

The mission was the third and final successful use of the Apollo spacecraft to take a crew to the Skylab space station and return them safely to earth.
(See also Appendix 1: Skylab Space Station)

Apollo-Soyuz Test Project
Launch vehicle: Saturn 1B
Launch: 15 July 1975
Recovery: 24 July 1975
Crew: Vance Brand, Donald Slayton, Tom Stafford

The purpose of the Apollo-Soyuz Test Project (ASTP) was to rendezvous dock an American Apollo spacecraft with a Soyuz spacecraft of the Soviet Union. The Apollo spacecraft was basically the same as the Apollo used on the lunar and Skylab missions, but a special docking module (DM) was designed and built to facilitate the docking of two very dissimilar spacecraft. The DM, designed and produced entirely in the United States, was launched with the Apollo. It was 10 feet 4 inches long, 4 feet 8 inches in diameter and weighed 4436 pounds. The DM was essentially an air lock with a docking apparatus on one end that was compatible with Apollo; the other end was of Soviet design and compatible with Soyuz. The DM contained life-support systems and a communications interlink for connecting the two spacecraft when docked. The DM, always controlled from Apollo, also contained an electric furnace for joint high-temperature crystallization and materials processing experiments.

The Apollo spacecraft was launched from Kennedy Space Center at 3:50 pm EDT on 15 July, 7 hours 30 minutes after Soyuz was launched from the Baikonur Cosmodrome. The docking took place 44 hours 35 minutes Ground Elapsed Time (GET) after the Apollo launch, and 52 hours 5 minutes after the Soyuz launch. Because of its greater maneuverability, Apollo was the active partner in the docking exercise. Astronaut Stafford and Cosmonaut Aleksei Leonov shook hands 3 hours after docking. After 45 hours 6 minutes of joint operations, the five spacemen, who had each spent up to 7 hours in the other spacecraft, returned to their respective vehicles. Apollo released Soyuz 15 hours 16 minutes later. A second docking took place 32 minutes later, but this was done simply as an exercise and the crews did not meet face to face again. At 102 hours 16 minutes GET into the mission and 5 hours 59 minutes after the second docking, the two spacecrafts separated for the last time. Following their final separation from Soyuz, the Apollo crew released the DM at 199 hours 23 minutes GET so that simultaneous doppler tracking from earth of the CSM and DM could measure fluctuations in

The ASTP Apollo capsule with docking module.

Above: The Apollo spacecraft photographed from the Soyuz spacecraft. Stafford and Leonov (*top*) in the hatchway connecting the two spacecraft.
Right: The Apollo 17 CSM viewed from the LM.

their respective orbits against the gravitational forces generated by the plates forming the earth's crust.

Soyuz returned to earth 40 hours after final separation, but Apollo continued for a duration of 217 hours 28 minutes. Upon re-entry, the Apollo's pressure-release valve was opened as usual to equalize pressure between the atmosphere of the CM and the earth's atmosphere. The reaction control system (RCS) rockets had failed to cut off and the highly poisonous nitrogen tetroxide oxydizer gas entered the CM through the open valves. Cabin oxygen flow was put on high and on landing the crew were able to put on their oxygen masks to tide them over until the US Navy recovery team opened the hatch.

The Apollo-Soyuz Test Project represented the final space flight of an Apollo spacecraft. Another Apollo spacecraft, which had been held in reserve for use as a possible rescue ship during the Skylab and ASTP programs, was never used.

Apollo P&F Subsatellite

Launch vehicle: Apollo spacecraft
Launch: 4 August 1971 (Apollo 15)
23 April 1972 (Apollo 16)
Total weight: 80 pounds
Diameter: 14 inches
Height: 31 inches
Shape: Hexagonal

There were two Apollo P&F (particles and fields) experimental spacecraft designed to be spring launched into lunar orbit from the Apollo spacecraft. They were designed to investigate the moon's mass and gravitational variations, particle composition of space near the moon and the interaction of the moon's magnetic field with that of the earth. The payload included a 1-watt S-band transponder and a magnetic-core memory unit with 49,152-bit capacity and a transmit rate of 128 bits per second. Solar arrays provided for 25 watts of solar power. The orbit of the first spacecraft decayed within six months, but the second remained in normal operation until 29 May 1977, when it crashed into the moon.

Applications Explorer Missions

This program encompassed scientific research projects similar to those included in the broad spectrum of activities of the earlier Explorer numbered series.
(See also Explorer)

Applications Explorer Mission A (AEM-A)
Launch vehicle: Scout D
Launch: 26 April 1978
Total weight: 296.1 pounds
Height: 63.7 inches (including antenna)

The spacecraft structure consisted of two major components: an instrument module containing a heat capacity mapping radiometer and supporting equipment, and the spacecraft base module, containing data handling, power supply, communications and command, and attitude-control subsystems.

Known as the Heat Capacity Mapping Mission (HCMM) spacecraft, this satellite was designed to use both visible and infrared observations of the earth to measure heat retention capacity in the fields of geology, plant stress, soil moisture and hydrology.

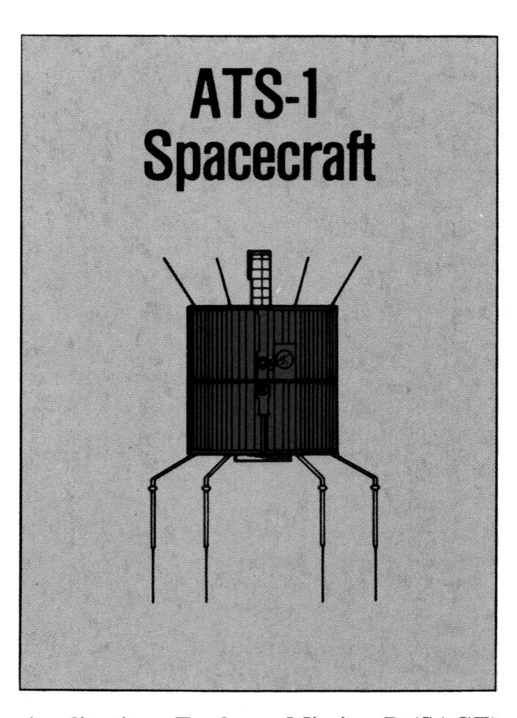

Applications Explorer Mission B (SAGE)
Launch vehicle: Scout
Launch: 18 February 1979
Diameter: 28 inches
Height: 25 inches (with two 24-square-foot solar arrays deployed winglike from the spacecraft)
Shape: Hexagonal

This program, known as the Stratospheric Aerosol and Gas Experiment (SAGE), was designed to measure solar radiation in the earth's atmospheric layers to establish baseline data on aerosol and ozone global concentrations to access the effects of transient phenomena on these concentrations.
(See also Explorer 55)

Applications Technology Satellites

The Applications Technology Satellites (ATS) were the first series of satellites to be placed in a geosynchronous orbit, which matched the rotation of the earth. This meant that they were 'parked' over the same spot on the earth throughout their operational life. Positioned at roughly 2047 miles above the earth, they could view about 45 percent of the earth's surface. The program was managed by NASA, with Hughes the primary spacecraft contractor for ATS-1 through ATS-5, and Fairchild the prime contractor for ATS-6.

ATS-1
Launch vehicle: Atlas Agena D
Launch: 7 December 1966
Total weight: 1550 pounds at launch
775 pounds in orbit
Diameter: 56 inches
Height: 57 inches
Shape: Cylindrical

Phased array experiment antennae and eight whip antennae extended from the top of ATS-1. Solar cells covered the sides. Eight VHF experiment antennae in the phased array and the apogee motor's nozzle projected from the satellite base.

ATS-1 was parked in a geosynchronous orbit over the Pacific near Hawaii, permitting color television transmissions between North America and the nations of the Pacific basin, such as Australia and Japan. The Spin Scan Cloud Camera returned the first photo covering nearly the entire disk of the earth on 9 December 1966. Spacecraft technology and scientific experiments were also in the payload.

ATS-2
Launch vehicle: Atlas Agena D
Launch: 6 April 1967
Re-entry: 2 September 1969
Total weight: 815 pounds
Diameter: 56 inches
Height: 72 inches
Shape: Cylindrical

The primary mission of the ATS-2 spacecraft was to test a gravity gradient control system in a 6900-mile orbit. A failure of the Agena second stage resulted in an elliptical rather than circular orbit, precluding meaningful evaluation of gravity gradient experiments, and resulted in limited data from 11 other experiments. These included experiments in communications and meteorology as well as a DOD albedo experiment.

Two pairs of 123-foot gravity gradient primary stabilization booms and a pair of 45-foot damping booms were deployed from the midsection, eight command and telemetry antennae extended from the top and four communication experiment antennae projected from the bottom. Solar cells covered the sides except for the midsection experiment band.

ATS-3
Launch vehicle: Atlas Agena D
Launch: 5 November 1967
Total weight: 805 pounds
Diameter: 5 feet
Height: 6 feet
Shape: Cylindrical

The sides of the ATS-3 were covered with solar cells except for a camera opening. The forward panel supported sun sensors, axial jets, and eight 5-foot-long VHF experiment whip antennae mounted around the forward end. A thrust tube extended from the aft end and contained the apogee kick motor.

This spacecraft carried nine experiments involving communications, meteorology, earth photography in color, navigation, stabilization and pointing, and degradation of surfaces in space and the ionosphere. It returned the first color photographs of the earth from geosynchronous orbit on 5 November 1967. ATS-3 was later moved to a different geosynchronous orbit to provide better views of the earth during the storm season.

ATS-4
Launch vehicle: Atlas Centaur
Launch: 10 August 1968
Re-entry: 17 October 1968
Total weight: 864 pounds after apogee motor burnout and jettison
Diameter: 56 inches
Height: 72 inches
Shape: Cylindrical

The exterior of the ATS-3 was covered with 21,864 n-on-p solar cells, except for an experiment band around the midsection, which contained a magnetometer sensor and ion and hydrazine gas jets. Appendages included eight quarter-wave whip antennae extended fanlike from the forward end, four motor-driven gravity-gradient booms mounted inside the experiment band, each extendable to 123 feet and forming a giant X when extended from the aft end.

The spacecraft performed communications, meteorology, technology and science experiments. However, the gravity-gradient experiments could not be conducted because the spacecraft did not separate from the Centaur, and failure of the latter to fire properly in space prevented ATS-4 from achieving its prover parking orbit, which in turn led to premature re-entry of the spacecraft.

ATS-5
Launch vehicle: Atlas Centaur
Launch: 12 August 1969
Total weight: 954 pounds
Description: Same as ATS-4

After a perfect launch, the spacecraft began to wobble on its spin axis during transfer to the parking orbit. Although 9 of the 13 experiments returned useful data, the primary gravity-gradient experiment aimed at providing information for stabilization control design of long-life spacecraft in geosynchronous orbit could not be performed because the counterclockwise spin prevented the experiment booms from being deployed.

ATS-1, the first of five built for NASA by Hughes.

ATS-6 (also called ATS-F)
Launch vehicle: Titan 3C
Launch: 30 May 1974
Total weight: 3090 pounds
Height: 26 feet

The ATS-6 was the most complex and powerful communications system in the 15-year history of communications satellites. It consisted of an earth-viewing module (EVM) housing controls and earth-viewing experiments. The EVM was attached to a 30-foot diameter deployable reflector antenna by a tubular support truss. Two structural arms, each supporting a semicylindrical solar array, extended from a hub that supported the reflector. The EVM was built in three separate sections. The bottom section was the experiments module that housed antennae and experiments that required earth-pointing orientation. The center section, the SM, contained attitude control, propulsion, telemetry and command, and parts of the power-supply subsystems. The upper section, the communications module, housed the transponder. Antenna feeds that radiated energy from the transponder to the 9-meter reflector for relay to earth were located on top of the communications module. The EVM weighed 2000 pounds, was 54 by 54 inches on the sides and was 63 inches high. The reflector support truss that connected the EVM to the 30-foot antenna was made of graphite-reinforced plastic. The truss was 13 feet high and 54 inches square. The deployable reflector consisted of a 5-foot aluminum hub from which protruded 48 aluminum ribs. The ribs were covered with a flexible mesh of copper-coated

ATS-5 is fitted to the Atlas-Centaur launch vehicle. The shroud is in the background.

dacron covered with silicon. The antennae weighed 180 pounds.

Successfully placed into an initial parking orbit over the equator off the west coast of South America, the ATS-6 was moved a year later to a parking orbit over Lake Victoria in Kenya. It served as an international broadcasting station in space, and was used to broadcast health and educational television to small ground stations in remote locations. It also provided aeronautical and maritime communications, position location, data relay and spacecraft tracking.

ASAT

General term for antisatellite weapon (See also Air Launched Miniature Vehicle, Calsphere, Squanto Terror)

Atmosphere Explorer

(See Explorers 27, 32, 51, 54, 55)

Beacon

Launch vehicle: Jupiter C (Juno)
Launch: 23 October 1958 (Beacon 1)
14 August 1959 (Beacon 2)
Re-entry: 23 October 1958 (Beacon 1)
14 August 1959 (Beacon 2)
Total weight: 9.26 pounds
Diameter: 12 feet (when inflated)
Shape: Spherical (when inflated)

The Beacon was contained in the 29.57-pound fourth stage of the Juno 1, which was 50 inches long and 7 inches in diameter. This stage contained a solid propellant rocket and the payload case. The payload case contained the folded Beacon, which was basically a balloon, a pressure bottle to fill the sphere, the sphere ejection system and a radio transmitter. The shell of the Beacon was composed of mylar polyester film coated with aluminum foil. The purpose of Beacon was to study atmosphere density at various altitudes during an estimated two-week lifetime, but both failed due to a launch vehicle upper-stage malfunction.

Beacon Explorer

(See Explorers S-66, 22, 27)

Big Bird

Launch vehicle: Titan 3D/Agena
First launch: June 1971
Total weight: 25,000 to 29,000 pounds (approx)
Diameter: 10 feet (approx)
Height: 4 feet 6 inches (approx)

Officially known as Broad Coverage Photo Reconnaissance satellites (Code 467), Big Birds are a successful series of classified spacecraft built for the US Air Force by Lockheed. These satellites have an apogee of about 150 miles, a perigee of 100 miles and a useful life of approximately six months. A specialized version of the spacecraft flies at lower altitudes, but for only about half the duration. Big Bird is equipped to conduct both television and radio surveillance, and it is also capable of taking photographs with high-resolution optical cameras. The film from these cameras is sent back to earth via recoverable 'film-return' capsules and the film is processed. The Big Bird reportedly can return as many as four of these capsules during its lifetime. The Big Bird is similar to the newer and more highly classified KH-11 reconnaissance satellite.

Biosatellite

Intended to study the effects of a space environment on living organisms, the biosatellite program ironically began five years after the start of American manned space travel. Of the planned six biosatellites, only three were launched before the program was canceled. The program was managed by NASA's Ames Research Center, and General Electric was the prime spacecraft contractor.

Biosatellite 1

Launch vehicle: Thrust Augmented Delta
Re-entry: 15 February 1967
Total weight: 940 pounds
Diameter: 40 to 57 inches (adapter section)
Height: 81 inches

Biosatellite 1 was composed of an adapter section and re-entry vehicle. The adapter section was a cylinder-cone and the re-entry vehicle was a blunt cone with a base 40 inches in diameter. It included a 280-pound aluminum blunt-cone experiment capsule with 6 cubic feet of payload space.

Seven radiation experiments were located in the forward section of the experiment capsule ahead of a Strontium 85 radiation source. Living organisms here included bacteria, bread mold, a flowering plant, flour beetles, parasitic wasps and common fruit flies (adult and larvae). Control versions of the same were located in a shielded section aft, along with six general biology experiments including amoebae, frog eggs, pepper plants and wheat seedlings. Life support systems provided for sea-level pressure, a temperature between 65 and 75 degrees Fahrenheit, and an oxygen-nitrogen atmosphere (like that of the earth but unlike the pure oxygen atmosphere of American manned spacecraft). A pair of 2-watt transmitters were aboard, one of which returned data to earth. Power was provided by silver-zinc batteries.

After a normal launch all systems worked as intended during the three-day mission. The re-entry vehicle separated normally on the 47th orbit, but the retrorocket failed to fire and the spacecraft could not return to earth as planned.

Biosatellite 2
Launch vehicle: Thrust Augmented Delta
Launch: 7 September 1967
Recovery: 9 September 1967
Total weight: 940 pounds
Height: 6 feet 9 inches

Biosatellite 2 was a cylinder-cone connected to a blunt-cone end. It consisted of three major segments: an adapter, a re-entry vehicle and a recovery capsule. The adapter was a 5-inch diameter cylinder whose forward end tapered to a 40-inch diameter opening, which in turn connected to the re-entry vehicle. The re-entry vehicle was a bowl-shaped heat shield 40 inches in diameter and 33 inches deep, with ablative surface material of phenolic nylon, an aft thermal cover, the thrust cone with retrorocket and cold gas spin and despin jets. The thrust cone assembly was fastened to the aft end of the heat shield to form the aft end of the re-entry vehicle. The recovery capsule was a covered aluminum bowl 31 inches in diameter with 6 cubic feet of payload space and a mounted inside heat shield. It contained an experimental payload together with associated support equipment.

There were 13 biology experiments arranged concentrically around a Strontium 85 radiation source isolate in a tungsten-nickel-copper sphere. The experiments were to test the effect of weightlessness on the changes produced in organisms by radiation. General biology

Above: Technicians check out Biosatellite 3, designed to carry a monkey into space.

Below: The pigtail monkey Bonny underwent nine days of continuous monitoring in space.

experiments and the control group for the radiation experiment were carried aft. The experimental organisms, the atmosphere and power source for Biosatellite 2 were the same as for Biosatellite 1.

The spacecraft re-entered properly and was recovered. The results of the experiments showed, among other things, that wheat seedlings grew 33 percent faster in a weightless environment, radiation damage is greater in weightlessness, the greatest effects of weightlessness (and radiation plus weightlessness) are on younger organisms, and that animal organisms are less affected than plant life.

Biosatellite 3
Launch vehicle: Thrust Augmented Delta N
Launch: 29 June 1969
Recovery: 7 July 1969
Total weight: 1536 pounds
Height: 7 feet
Diameter: 57 inches (adapter cylinder)
40 inches (heat shield)
Shape: Truncated cone

Biosatellite 3 was composed of three segments: an adapter, a re-entry vehicle and a recovery capsule. The power system, atmosphere and general characteristics are similar to Biosatellites 1 and 2.

Where Biosatellites 1 and 2 had carried similar, less complex organisms, this spacecraft was designed to carry a male pigtail monkey (*Macaca nemestrina*) to determine the effects of the space environment on mental, emotional and physiological processes of higher primates. Eleven sensors were attached to his brain, two to his eyes, two to his neck and back muscles and one to the bladder. Meanwhile, electrocardiographic (EKG) and blood-pressure data were checked. The monkey, named Bonny, had been trained to operate the feeding system manually. The mission, intended to last 30 days, was launched successfully, but the monkey reacted badly to weightlessness and began refusing to eat and drink. The flight was terminated after nine days, with 80 percent of the expected data obtained and with the monkey still alive. He died eight hours later of a heart attack, judged to have been brought about by dehydration.

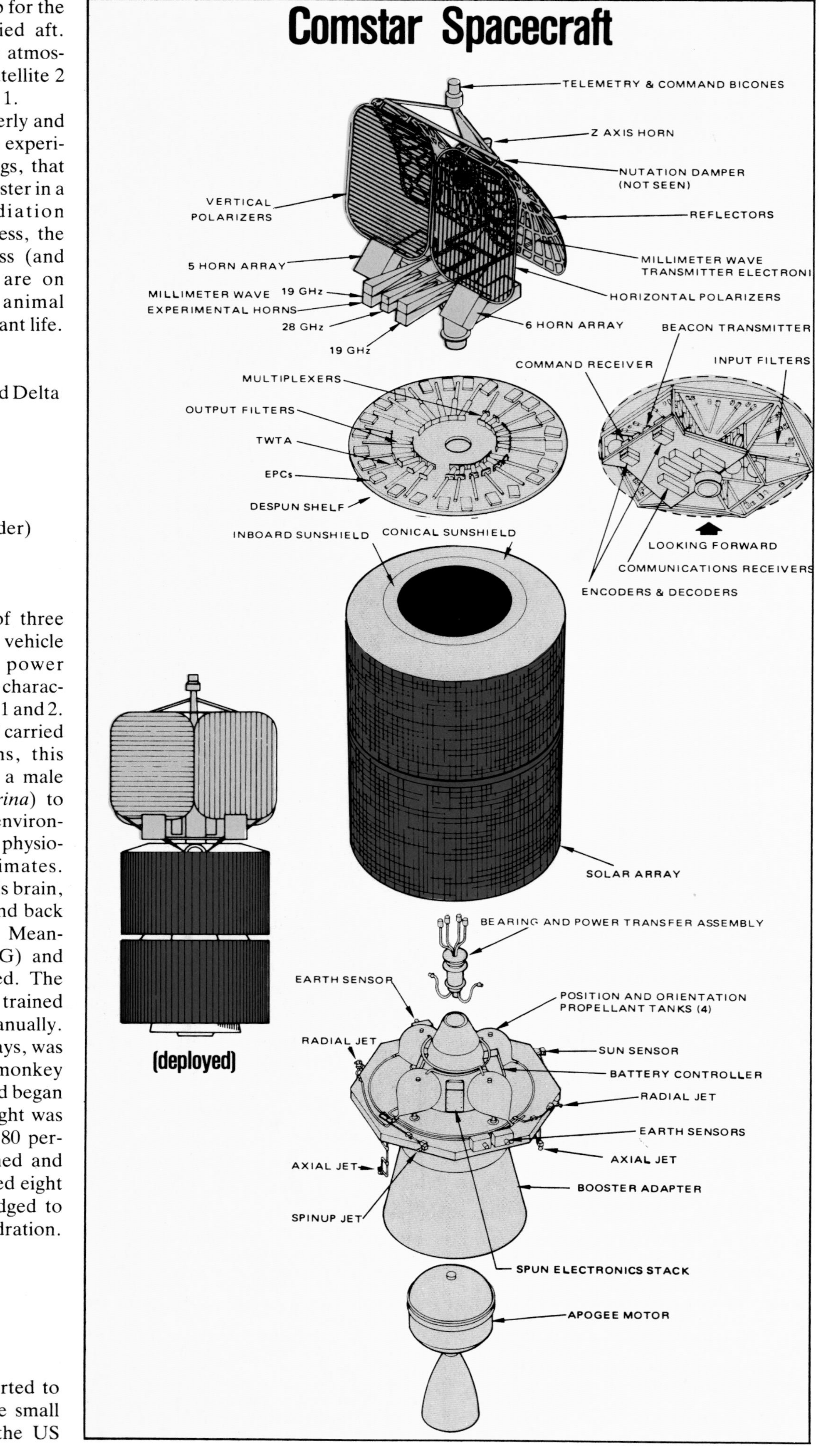

Calsphere

Launch vehicle: Thor/Burner 2
First launch: 6 October 1964

The Calsphere program is reported to have been a series of at least five small spacecraft developed jointly by the US

Air Force and US Navy to be used to destroy enemy satellites. Though details remain classified, at least one successful test is reported to have taken place.

Cameo

(See Nimbus 7)

Combined Radiation and Release Satellite

Launch vehicle: Space shuttle
Launch: 1985-86
Total weight: 9500 pounds

The Combined Radiation and Release Satellite (CRRS) is being developed by Ball Aerospace for NASA and the US Air Force. For the first 60 days of operation, it will conduct NASA chemical release experiment in low-earth orbit, and then conduct an Air Force radiation-effects measurement mission.

COMSAT

The Communications Satellite Corporation (COMSAT) is the owner and operator of the Comstar and Intelsat spacecraft.
(See Comstar, Intelsat, Marisat)

Comstar

The Comstar spacecraft were domestic telephone communications satellites designed by Hughes Aircraft for COMSAT General Corporation to provide easy communications between the contiguous United States, Alaska, Hawaii and Puerto Rico. Principal use of these satellites was shared by American Telephone and Telegraph (AT&T) and General Telephone and Electronics (GTE), who owned the seven ground stations in the system. The spacecraft were launched by NASA from Kennedy Space Center. There were four Comstars, and their designation system changed from numerical to alphabetical midway through the program. Original designations are given first.

Comstar 1 (Comstar A), Comstar 2 (Comstar B)
Launch vehicle: Atlas Centaur
Launch: 13 May 1976 (Comstar 1)
22 July 1976 (Comstar 2)
Total weight: 3300 pounds at launch
1800 pounds in orbit
Diameter: 8 feet
Height: 20 feet (including 9-foot 6-inch antennae on top of the main body)

Comstar 1 was leased to AT&T by COMSAT.

The main body was cylindrical. Solar arrays around the cylinder provided more than 600 watts of power. An apogee motor mounted in the base of the cylinder drove the spacecraft to its final orbit. A total of 24 transponder communications systems operating in the 4 to 6 GHz range provide 14,400 telephone circuits (per spacecraft).

Comstar D-3 (Comstar 3, or Comstar C)
Launch vehicle: Atlas Centaur
Launch: 29 June 1978
Total weight: 3342 pounds at launch
1746 pounds in orbit
Diameter: 8 feet
Height: 19 feet
Shape: Cylindrical

Comstar D-3 was a spin-stabilized satellite. Cylindrical solar panels, covered with nearly 17,000 solar cells, provided primary power of 760 watts. It had a design life of seven years. The payload was the same as for Comstars 1 and 2, but Comstar 3 had an operational receive frequency band in the 5.9 to 6.4 GHz range and a transmit frequency band in the 3.7 to 4.2 GHz range. An experimental package also tested communications in the 19 and 29 GHz range.

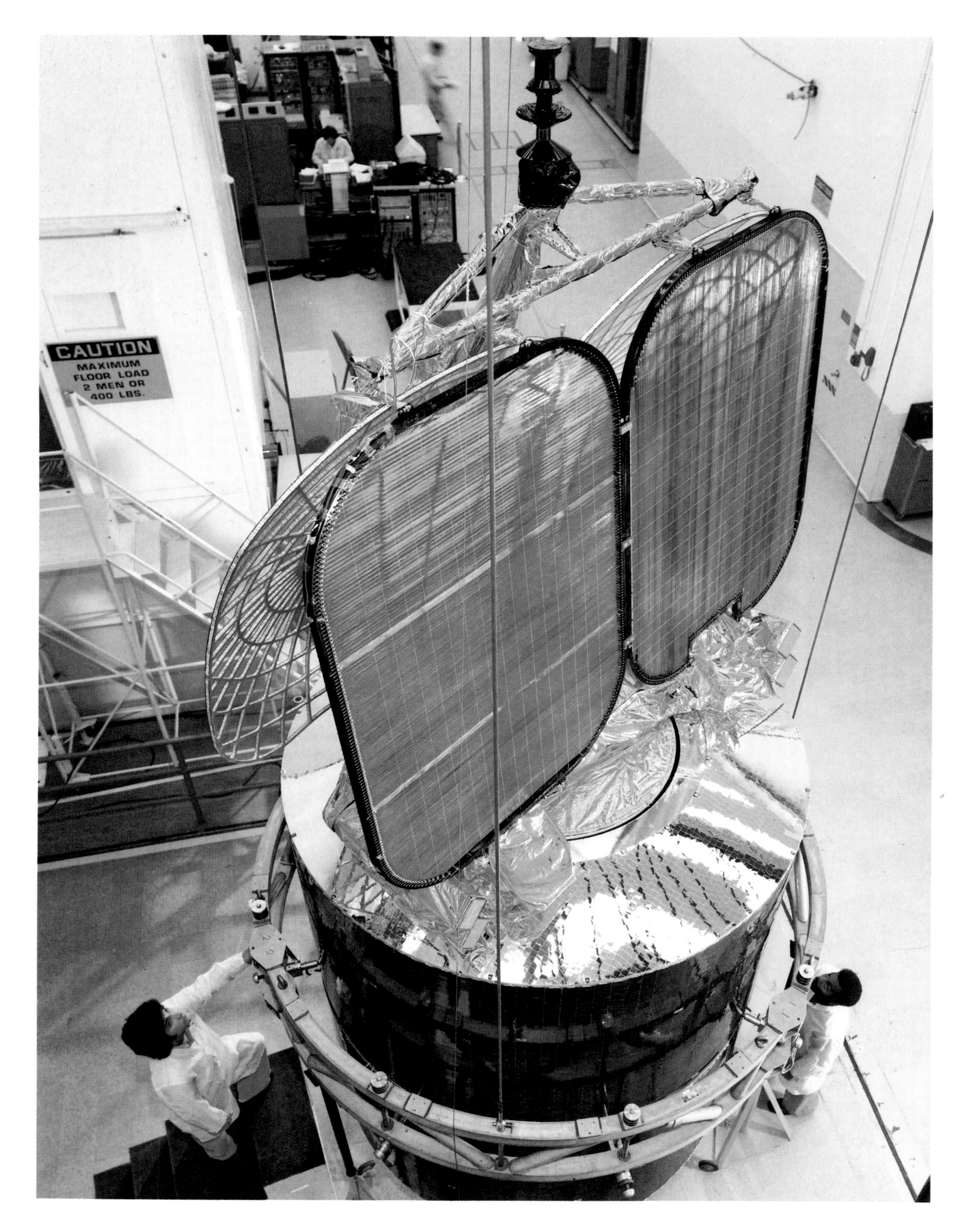
CAUTION
MAXIMUM
FLOOR LOAD
2 MEN OR
400 LBS.

Comstar D (Comstar 4)
Launch vehicle: Atlas Centaur
Launch: 21 February 1981
Total weight: 3342 pounds
Height: 20 feet
Shape: Cylindrical

At launch, the Comstar D carried 12 transponders, or channels, each capable of relaying 1500 one-way voice circuits and providing an overall communications capability of 18,000 simultaneous high-quality, two-way telephone transmissions. The payload was similar to earlier Comstars, but capable of 18,000 simultaneous circuits. Receive and transmit frequency bands were the same as Comstar 3, but experimental frequencies were at 19 and 28 GHz.

Cosmic Background Explorer

Launch vehicle: Space shuttle
Launch: September 1985
Total weight: 10,000 pounds

This Cosmic Background Explorer (COBE) spacecraft was designed by NASA's Goddard Space Flight Center to measure residual radiation from the 'Big Bang,' the event credited by some as being responsible for the present format of the universe. Its orbit altitude of 562.5 miles is at a 99-degree inclination.

Courier

Launch vehicle: Atlas B
Launch: 4 October 1960 (Courier 1B)
Last transmission: 21 October 1960
Total weight: 500 pounds

This US Army communications satellite succeeded SCORE as the second such spacecraft.
(See also DSCS, FLTSATCOM, SCORE)

CRL

Launch vehicle: Scout
Launch: 28 June 1963
Total weight: 259.6 pounds

This US Air Force program involved a Cambridge Research Laboratory (CRL) spacecraft designed for geophysics research and launched by NASA on a reimbursable basis.

Rectangular antenna reflectors on the Comstar allow the same frequency to be used twice.

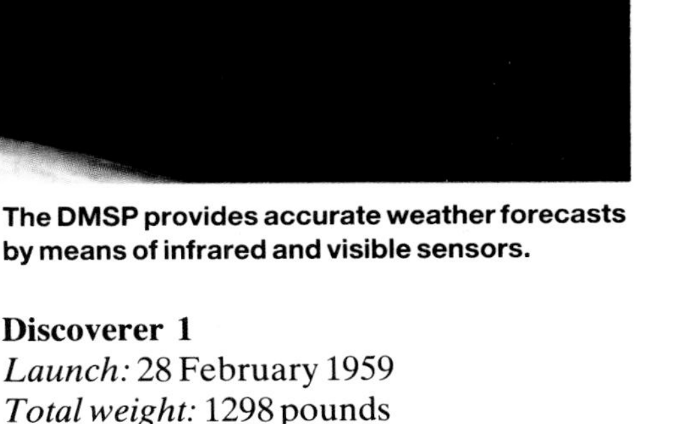

The DMSP provides accurate weather forecasts by means of infrared and visible sensors.

DAD

Launch vehicle: Scout
Launch and re-entry: 5 December 1975
Total weight: 88 pounds (DAD-A)
94.6 pounds (DAD-B)

These scientific spacecraft were designed to measure global density of the upper atmosphere and lower exosphere. They were launched from the Western Space and Missile Center, but the launch vehicle failed and orbit was not achieved.

DBSC

Launch vehicle: Space shuttle
Launch: 1987
Total weight: 3500 pounds

Designed by Ford Aerospace, this spacecraft will provide direct broadcast television services.

Discoverer

The Discoverer series spacecraft constituted the first major US military satellite program. Although the program had teething troubles, these spacecraft were extremely valuable in the development of later film-return-type reconnaissance satellites. Launched from the US Air Force's Western Space and Missile Center at Vandenberg AFB with Thor Agena launch vehicles, there were 38 spacecraft in the series. Similar satellites were launched subsequent to Discoverer 38 in 1962, but after this year military launches were classified, and there were no more spacecraft with official Discoverer designations. Although many of these satellites were lost owing to launch vehicle failure, their lifespan, even when successful, was shorter than a month. The following are some program highlights.

Discoverer 1
Launch: 28 February 1959
Total weight: 1298 pounds

This polar orbit test satellite could not be tracked due to tumbling.

Discoverer 2
Launch: 13 April 1959
Total weight: 1738 pounds (including 194-pound first returnable capsule)

The capsule of Discoverer 2 ejected on the 17th orbit, but was lost in the Arctic.

Discoverer 14
Launch: 10 August 1960

This was the first successful recovery of a film-return capsule after successful ejection. Recovery had been attempted on Discoverers 5, 6, 8 and 11, but failed.

Discoverer 38
Launch: 27 February 1962
Total weight: 2094 pounds

The capsule of this last Discoverer spacecraft was successfully recovered in midair on the 65th orbit (the 13th recovery, the ninth in midair).

DMSP

Launch vehicle: Atlas E
First launch: October 1971 (Block 5D)
November 1983 (Block 5D-2)
Total weight: 1043 pounds (Block 5D)
1131 pounds (Block 5D-1)
1650 pounds (Block 5D-2)
Height: 17 feet
14 feet (Block 5D-2)
Diameter: 6 feet
5 feet (Block 5D-2)

The DMSP (Defense Meteorological Satellite Program) spacecraft are composed of three parts, including a 400-pound precision sensor platform (seven sensors) on top, an electronics platform and a third-stage rocket enclosed in an altitude-control system at the base. The one steerable solar array generates 900 watts of power. Four pitch/yaw jets provide maneuverability.

Designed and built by RCA, the DMSP spacecraft replace earlier RCA satellites as the primary military meteorological reconnaissance satellites. Similar to NOAA's TIROS spacecraft, the DMSP satellites transmit weather data to the US Air Force Global Weather Central at Strategic Air Command Headquarters (Offutt AFB, Nebraska) and to other American military installations worldwide. The spacecraft's sensors sweep an area 1800 miles wide with resolutions down to 1800 feet. The sensors include an electron spectrometer to evaluate atmospheric conditions that might effect communications, a radiometer to measure vertical temperature of water vapor in the atmosphere, a microwave imager to gather weather data through cloud cover, a microwave temperature gauge to measure temperature from ground level to 18 miles' altitude, and an ultraviolet sensor to measure atmospheric density.

DSCS

The DSCS (Defense Satellite Communications System), referred to as 'Discus,' consists of three series, or phases, of spacecraft designed to provide DOD with a reliable network of strategic communications satellites with global coverage. Managed by the US Air Force, the DSCS satellites were developed by TRW (Thompson Ramo Wooldridge, Inc), the primary spacecraft contractor on the DSCS 1/DSCS 2 program, and by General Electric, the spacecraft contractor for DSCS 3.

DSCS 1 (Initial DSCS, or IDSCS)
Launch vehicle: Titan 3C
First launch: 16 June 1966
Last launch: 13 June 1968
Total weight: 99 pounds
Diameter: 33.5 inches
Surface composition: Solar cells
Shape: 26-sided polygon

The first phase of the DSCS program involved the launch of 26 spacecraft offering single-channel voice relay, imagery, teletype and computerized digital data transmission. Though they were designed with intended lifespans of only 18 months, one of the original DSCS 1 spacecraft was still operating in 1982.

DSCS 2
Launch vehicle: Titan 3C or 34D/IUS
First launch: 3 November 1971
Total weight: 1245 pounds
Height: 6 feet (13 feet with sensors deployed)
Diameter: 9 feet
Shape: Cylindrical

The second phase of DSCS involved spacecraft designed with five-year lifespans placed in 23,230 nautical-mile geosynchronous orbits with earth coverage and spot beam antennae on each, providing up to 1300 duplex voice communication channels on the 14 and 16 GHz frequencies. There were 16 DSCS 2 spacecraft delivered, and the first six were launched in pairs in November 1971, December 1971, December 1973 and May 1975. Further launches took place later in the decade, and four spacecraft were lost due to launch vehicle failure. Seven were still operational as of 1982, the launch year of a modified DSCS 2.

DSCS 3
Launch vehicles: Titan 34C space shuttle/ IUS
First launch: 30 October 1982
Total weight: 1876 pounds

Operating in the same geosynchronous equatorial orbit as its predecessor, DSCS 3 offers 50 percent greater communications capability and has a 10-year life span. Like DSCS 2, it is controlled by the US Air Force Satellite Control Facility near San Francisco, with communications control built into selected Defense Communications Agency terminals. Through its larger antenna array and steerable beams, DSCS 3 provides earth coverage as well as spot-beam transmissions to smaller portable ground stations.

DSPS

Launch vehicle: Titan 3C
First launch: 5 May 1971
Total weight: 2000 pounds

Above: DSCS 3 has increased antijam protection and communications capability over DSCS 2 (*left*). *Below:* Dynamics Explorer-A is shown undergoing preflight evaluation testing at GSFC, Maryland.

Above: The latest USAF/TRW DSPS was launched by the space shuttle in January 1985.

Placed in geosynchronous orbit, the DSPSs, or Defense Support Program Satellites (Code 647), use infrared sensors to detect launches of Intercontinental Ballistic Missiles (ICBMs) and Submarine-Launched Ballistic Missiles (SLBMs).

Dynamics Explorer

Launch vehicle: Delta
Launch: 3 August 1981
Total weight: 888 pounds (A)
915 pounds (B)
Height: 45 inches (each)
Diameter: 53 inches (each)
Shape: 16-sided polygonal

The Dynamics Explorer spacecraft were two very similar satellites launched simultaneously from the Western Space and Missile Center to study space around the earth from the fringe of the upper atmosphere to distances outside the earth's magnetic field. The DE-A (or DE-1) spacecraft was placed in a 15,000-mile orbit, while DE-B (DE-2) was placed in 1400-mile polar orbit. Though the orbits were 10 to 23 percent shallower than planned, the RCA-produced spacecraft were able to carry out their missions.

Early Bird

(See Intelsat 1)

Echo

This series of spherical satellites was designed as passive communications satellites to reflect, or 'bounce,' radio communications, hence the name 'Echo.' Echo A 10, launched from Cape Canaveral, was lost because of second-stage malfunction. Echo 1, which was also launched from Cape Canaveral, was successful and helped transmit the first two-way voice communication and television

transmission via satellite. Echo AVT-1 and Echo AVT-2 were successfully launched, but the AVT-1 sphere ruptured and AVT-2 did not inflate properly. Echo 2, successfully launched from the Western Space and Missile Center, is thought to have not inflated properly, but good experiment results were obtained nevertheless.

Above: Echo 1 was used for communications experiments for almost eight years.

Below: The folded Echo 2 is placed inside the adapter canister.

Echo A 10
Launch vehicle: Thor Delta
Launch and re-entry: 13 May 1960
Total weight: 166 pounds
Diameter: 100 feet
Shell composition: Mylar polyester film, covered with vapor-deposited aluminum
Shape: Spherical

Echo 1 (A 11)
Launch vehicle: Thor Delta
Launch: 12 August 1960
Re-entry: 24 May 1968
Total weight: 166 pounds
Diameter: 100 feet
Shell composition: Aluminum coated
Shape: Spherical balloon

Echo AVT
Launch vehicle: Thor
Launch: 15 January 1962 (AVT-1)
18 July 1962 (AVT-2)
Re-entry: 15 January 1962 (AVT-1)
18 July 1962 (AVT-2)
Diameter: 135 feet
Shape: Spherical

Above: This time-lapse exposure shows the Echo 1's diagonal trail against the Milky Way.

Below: Echo 2's canister is shown during drop tests in a vacuum chamber.

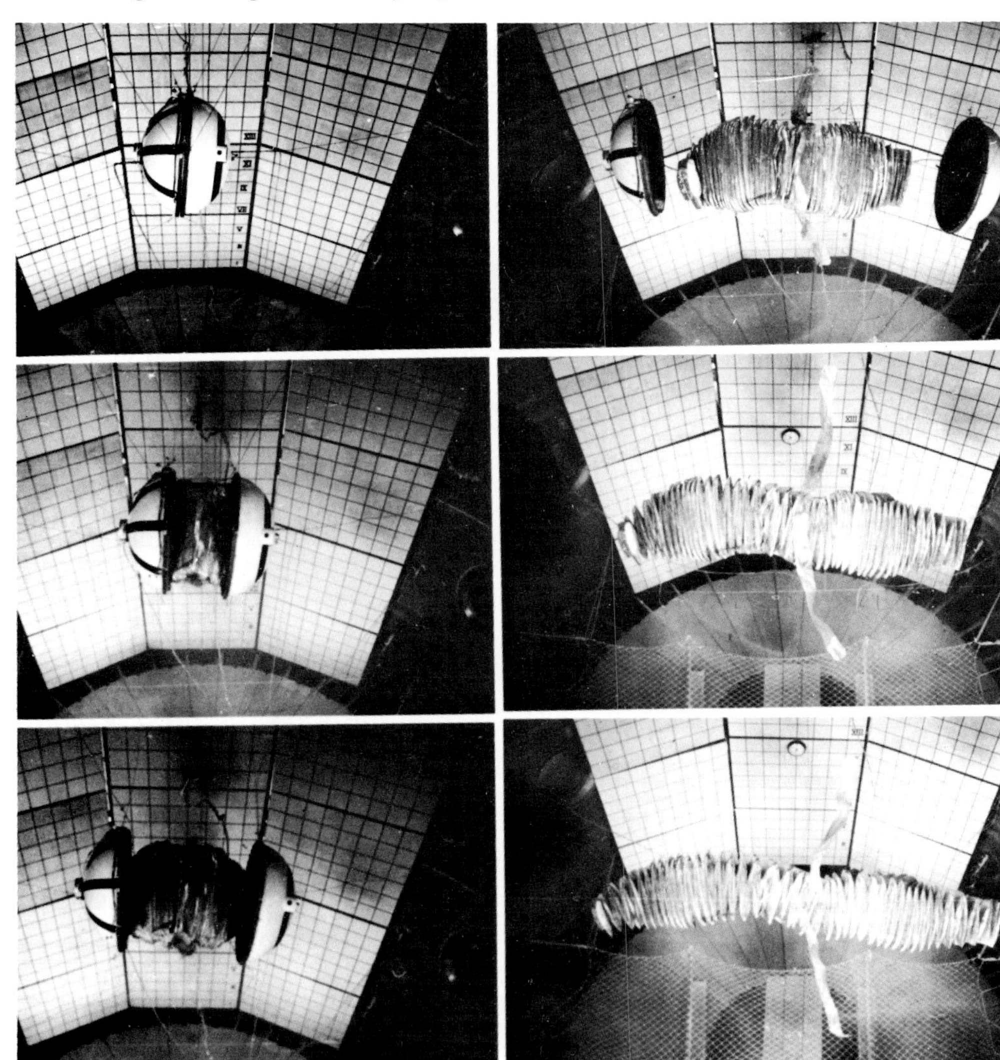

Echo 2
Launch vehicle: Thor Agena
Launch: 25 January 1964
Re-entry: 7 June 1969
Total weight: 457 pounds
Diameter: 135 feet
Shell composition: Mylar and aluminum alloy foil laminated to either side and coated on both sides with alodine
Shape: Spherical

ELINT

The ELINT spacecraft are a general variety of classified military satellites whose purpose is to gather electronic intelligence. They are used to monitor radar and radio signals originating on the ground, from aircraft and from other spacecraft. The data is used to keep track of the operations of potential enemies as well as to develop means of jamming enemy radar and radio, and to develop electronic countermeasures (ECMs) to counteract enemy jamming capability.

ELINT 1
First launch: March 1962
Total weight: 2000 pounds

ELINT 2
First launch: March 1963
Total weight: 3300 pounds

ELINT 3
First launch: 1968
Last launch: 1970
Total weight: 4400 pounds

ELINT 'ferret' satellite series
Launch vehicle: Designed to 'hitchhike' into space aboard larger reconnaissance spacecraft such as Big Bird
Total weight: 120 pounds

(See also Ferret)

ERBS

Launch vehicle: Space shuttle
Launch: 5 October 1984
Total weight: 5000 pounds
Diameter: 5 feet
Height: 12 feet 6 inches
Width: 15 feet (base module)

The earth radiation budget satellite, known as ERBS, is an important part of the US National Climate Control Program. It was released into orbit by astronaut Sally Ride, who operated the manipulator arm of the space shuttle orbiter *Challenger* on STS flight 41G.

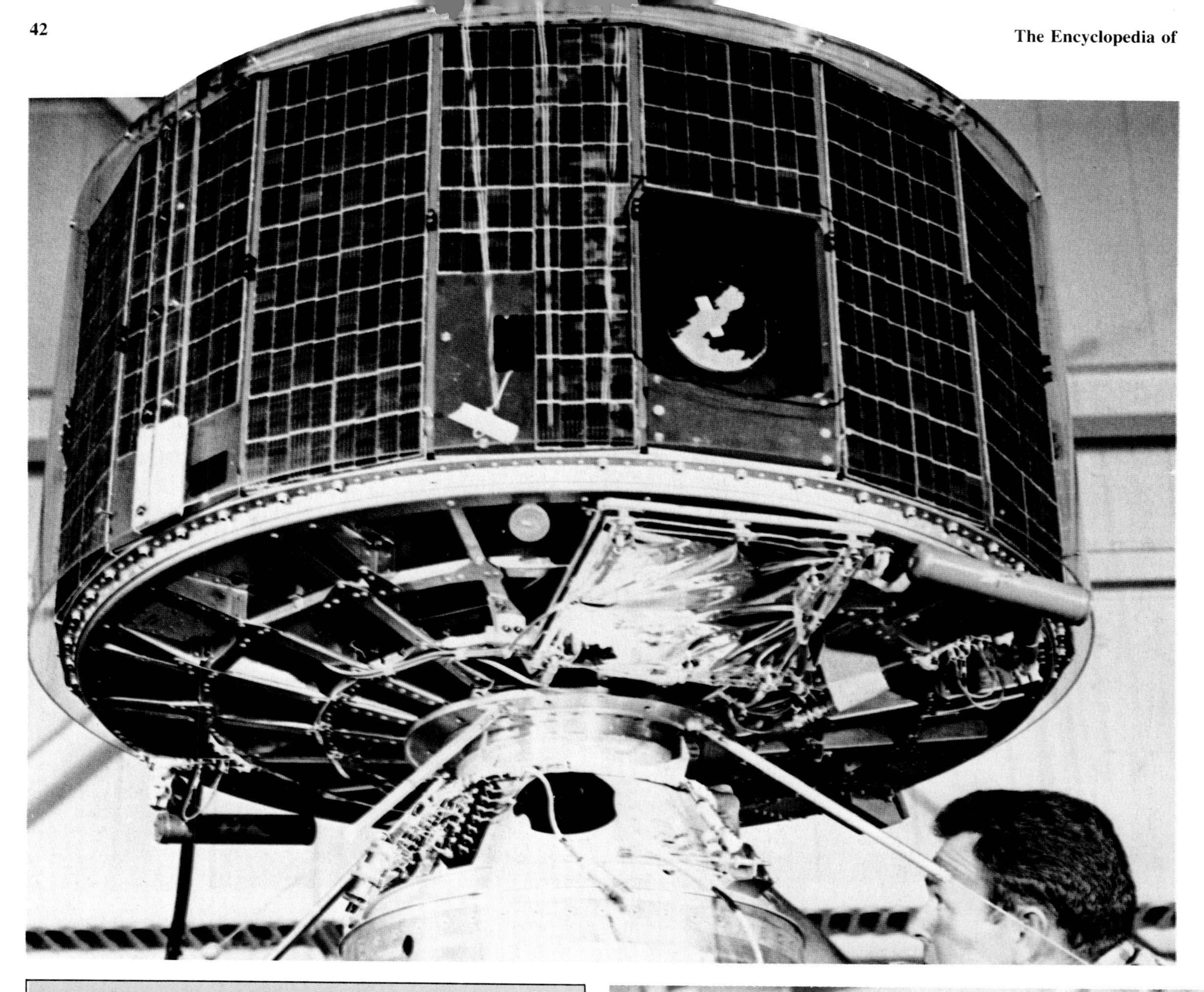

ESSA/TOS

Spacecraft	Launch	Duration (days)	Weight (lb)	Camera
ESSA 1	3 Feb 1966	861	305	TV
ESSA 2	28 Feb 1966	1692	290	APT
ESSA 3 (TOS-A)	2 Oct 1966	738	320	AVCS
ESSA 4 (TOS-B)	26 Jan 1967	465	290	APT
ESSA 5 (TOS-C)	20 April 1967	1034	320	AVCS
ESSA 6 (TOS-D)	10 Nov 1967	763	290	APT
ESSA 7 (TOS-E)	16 Aug 1968	576	320	AVCS
ESSA 8 (TOS-F)	15 Feb 1968	2644	290	APT
ESSA 9 (TOS-G)	26 Feb 1969	1726	320	AVCS

(See ITOS, NOAA, TIROS)

ESSA 3, the first TOS satellite to carry two AVCSs, is readied for launch (*left*). ESSA 5 (*below*) undergoes prelaunch inspection.

Above: ESSA 6 (TOS D) inside the shroud. It photographed a section of the earth's cloud cover every 6 minutes as it passed in orbit.

After initial deployment using the arm, the satellite fired its four hydrazine rocket motors, boosting it into a 330-nautical-mile orbit from where it would measure solar radiation reaching the earth. The ERBS climate research spacecraft was a follow-on to the NOAA spacecraft series and was designed to operate in conjunction with the NOAA-F and NOAA-G spacecraft launched in 1984 and 1985, respectively.

The spacecraft was built by Ball Aerospace and the Earth Radiation Budget experiment instruments were developed by TRW. The spacecraft also carries a Ball Aerospace SAGE-2 instrument. Launched by NASA, the program is managed by NASA's Goddard Spaceflight Center.

(See also Applications Explorer Missions, NOAA)

ERTS

(See Landsat)

ESSA

These meteorological spacecraft, similar in configuration to the Tiros spacecraft, were developed by NASA's Goddard Spaceflight Center and built by RCA for the US Environmental Science Services Administration (ESSA). The objective of the program, accomplished with ESSA 1 and 2, was to establish a global weather satellite system. From ESSA 2 on, these spacecraft carried either the automatic picture transmission (APT) systems or the advanced vidicom camera system (AVCS). These systems included two 800-line cameras covering a four-million-square-mile area at a time with resolution down to two miles. The first two spacecraft were launched from Cape Canaveral, but from ESSA 3 on, these spacecraft were launched from the Western Space and Missile Center at Vandenberg AFB and were also known as TIROS Operational Satellites (TOS).

ESSA

General specifications: (See table page 42)
Launch vehicle: Delta
Height: 22 inches
Diameter: 42 inches
Shape: Cylindrical 18-sided polygon

Two television cameras were mounted 180 degrees apart on the sides, an 18-inch receiving antenna was mounted on top, solar cells covered the top and sides, and four whip antennae extended from the baseplates of the ESSA.

(See also TIROS)

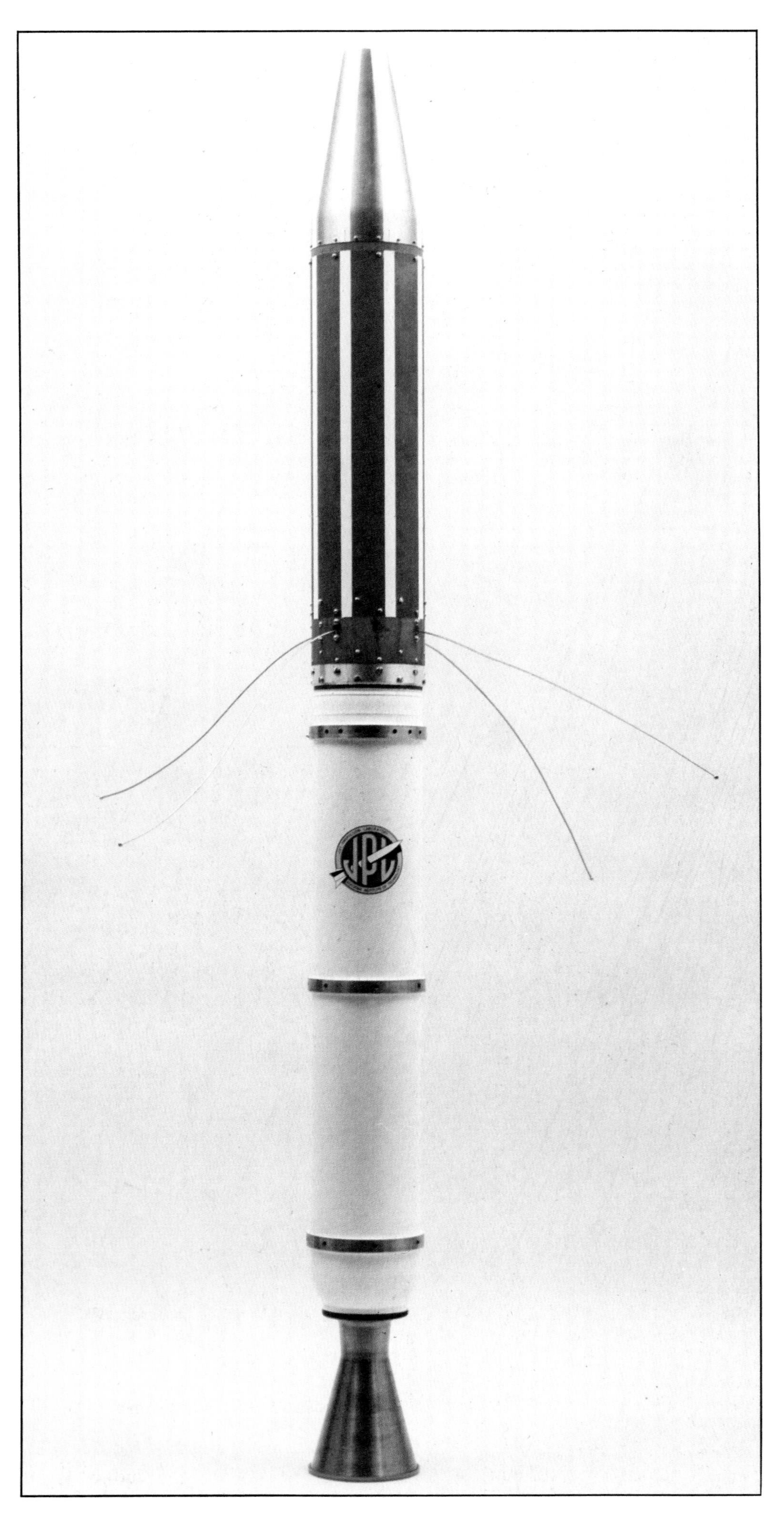

Explorer

The Explorer spacecraft program has been the most extensive in the history of American space exploration. Rivaled in size only by the Soviet Cosmos program, the Explorer program started before Cosmos and spanned nearly three decades. All of the Explorer satellites were designed for scientific research and built by NASA or other research organizations. All of them were managed by NASA except for the first five, which were managed by the Defense Advanced Research Projects Agency (ARPA, later DARPA). The launch vehicles were developed by Dr Wernher Von Braun of the Army Ballistic Missile Agency (ABMA). During World War II, von Braun developed Germany's V-2 (A-4) program, the first successful ballistic missile program. Launch of the first Explorer satellite came during the International Geophysical Year and four months after the launch of the world's first successful spacecraft, Russia's Sputnik 1, on 4 October 1957.

The Explorer spacecraft are listed below in chronological order. Those Explorers with *S-prefix* designations should not be confused with the other Explorers that might coincidentally have the same *number*. Because of their relationship to the rest of the Explorer program, the Explorer S spacecraft are integrated into the chronological list of Explorer spacecraft below. Explorer S-1 follows Explorer 5; Explorer S-2 *is* Explorer 6; Explorer S-45 follows Explorer 9; Explorer S-45a follows Explorer 11; Explorer S-46 follows Explorer 7; Explorer S-55 follows Explorer 11; Explorer S-55a, b and c *were* Explorer 13, 16 and 23; Explorer S-56 follows Explorer 8 and Explorer S-66 follows Explorer 17.

Explorer 1
Launch vehicle: Jupiter C
Launch: 31 January 1958
Re-entry: 31 March 1970
Total weight: 30.8 pounds
Length: 6 feet 8 inches

The first American satellite to achieve orbit, Explorer 1 completed 58,000 orbits and stayed in space for 12 years. Designed to measure and transmit data on temperature, meteorites and cosmic rays, the spacecraft confirmed the existence of the Van Allen radiation belt 600 miles above the earth. Explorer 1 continued to transmit data until 23 May 1958.

Explorer 1 was America's first spacecraft.

Explorer 2
Launch vehicle: Jupiter C
Launch and re-entry: 5 March 1958
Total weight: 30.8 pounds

Orbit was not achieved because of launch vehicle failure.

Explorer 3
Launch vehicle: Jupiter C
Launch: 26 March 1958
Re-entry: 28 June 1959
Total weight: 30.8 pounds

This spacecraft continued to store and transmit data on cosmic radiation to ground stations until 16 June 1958. It was similar to the earlier Explorers.

Explorer 4
Launch vehicle: Jupiter C
Launch: 26 July 1958
Re-entry: 23 October 1959
Total weight: 37.4 pounds

This spacecraft was designed to observe radiation from Project Argus and it transmitted until 23 October 1959.

Explorer 5
Launch vehicle: Jupiter C
Launch and re-entry: 24 August 1958
Total weight: 37.4 pounds

Orbit was not achieved because of launch vehicle failure.

Explorer S-1
Launch vehicle: Juno 2
Launch: 16 July 1959
Destroyed: 16 July 1959
Total weight: 91.5 pounds
Diameter: 30 inches
Height: 28 inches

The first NASA Explorer, this satellite was intended for much the same type of mission as the earlier ARPA Explorers. It consisted of two truncated cones joined at their bases and was purposely destroyed 5.5 seconds after lift-off when the launch vehicle went off course.

Explorer 6 (S-2)
Launch vehicle: Thor Able
Launch: 7 August 1959
Total weight: 142 pounds
Diameter: 26 inches
Height: 29 inches
Shell composition: Aluminum alloy

This satellite was an irregular spheroid with a slightly flattened bottom. Successfully placed in a better than expected orbit, Explorer 6 studied radiation levels, geomagnetic fields and the Van Allen Belt, and it returned the first televised cloud-cover pictures from space.

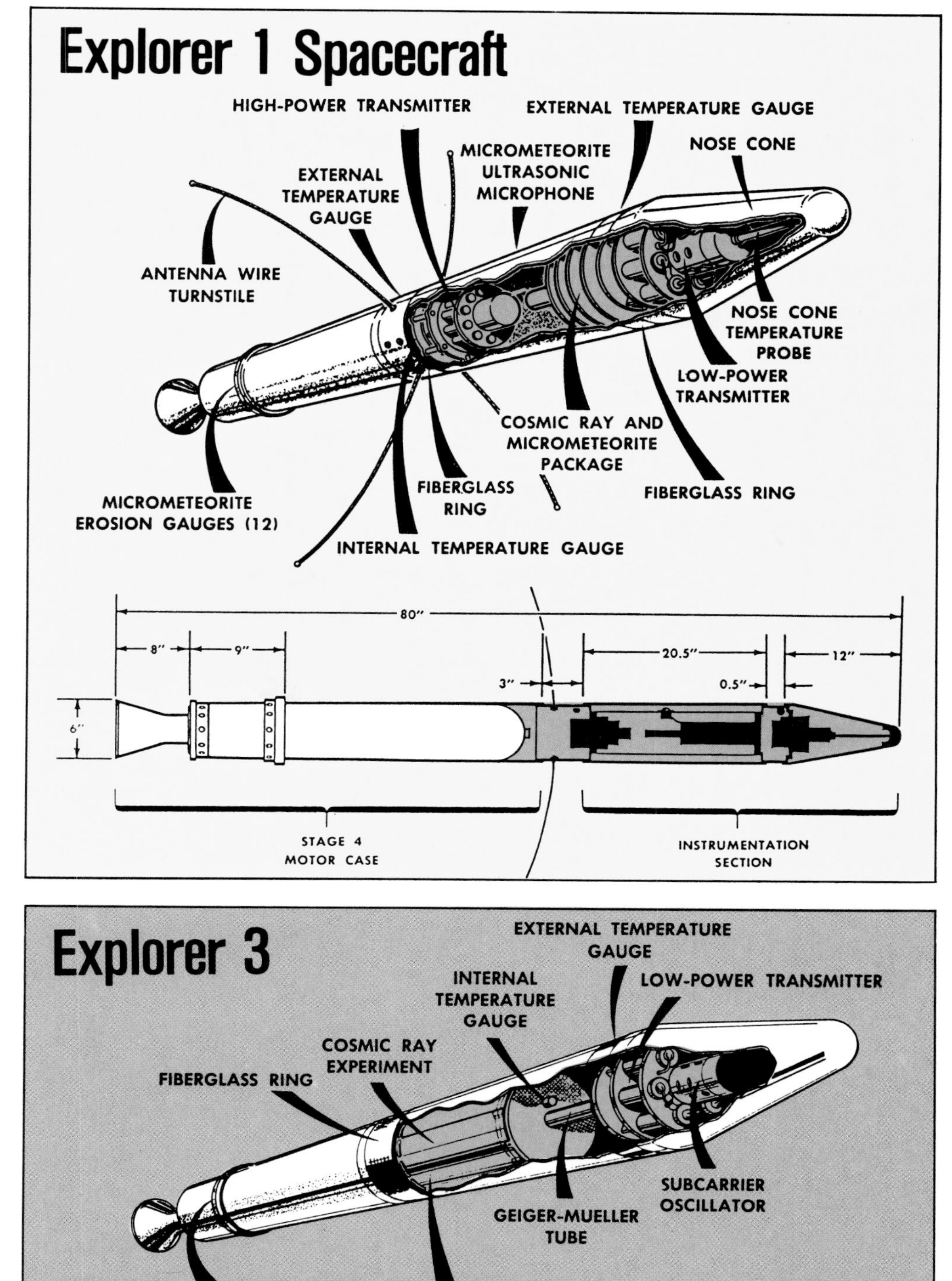

Explorer 7
Launch vehicle: Juno 2
Launch: 13 October 1959
Total weight: 91.5 pounds
Diameter: 30 inches
Height: 30 inches
Shell composition: Fiberglass and sandblasted aluminum foil
Shape: Two truncated cones joined at bases

Explorer 7 returned data on radiation and magnetic storms.

Explorer S-46
Launch vehicle: Juno 2
Launch and re-entry: 23 March 1960
Total weight: 35.3 pounds
Diameter: 7 inches
Height: 21 inches
Shell composition: Aluminum alloy
Shape: Cylindrical

Orbit was not achieved due to launch vehicle failure.

Explorer 8
Launch vehicle: Juno 2
Launch: 3 November 1960
Last transmission: 28 December 1960
Total weight: 90 pounds
Diameter: 30 inches
Height: 30 inches
Shell composition: Aluminum
Shape: Two truncated cones joined at bases

Explorer 8 confirmed the existence of a helium layer in the ionosphere.

Explorer S-56
Launch vehicle: Scout (first operational use)
Launch and re-entry: 4 December 1960
Total weight: 14 pounds
Diameter: 12 feet
Shell composition: Mylar polyester and aluminum foil
Shape: Spherical

Explorer S-56 was designed for atmospheric physics and launch vehicle testing, but orbit was not achieved due to launch vehicle failure.

Explorer 9
Launch vehicle: Scout (first successful operational use)
Launch: 16 February 1961
Re-entry: 9 April 1964
Total weight: 15 pounds
Diameter: 12 feet
Shape: Spherical

This was the first US spacecraft successfully placed into orbit with a solid-fuel launch vehicle. Intended to return data on atmospheric physics, the spacecraft's transmitter failed on the first orbit.

Explorer S-45
Launch vehicle: Juno 2
Launch and re-entry: 24 February 1961
Total weight: 74 pounds
Diameter: 30 inches
Height: 24 inches
Shell composition: Aluminum
Shape: Two truncated cones joined at the bases

Ionospheric research was not possible when orbit was not achieved due to launch vehicle failure.

Explorer 10 (P-14)
Launch vehicle: Thor Delta
Launch: 25 March 1961
Re-entry: June 1968
Total weight: 79 pounds
Height: 52 inches

Explorer 10 consisted of a 13-inch sphere atop a conical tube joining it to the top of a 10-inch drum. It returned 60 hours of data on the interplanetary magnetic field between the earth and sun (solar wind) and on solar flares.

Explorer 11
Launch vehicle: Four-stage Juno 2
Launch: 27 April 1961
Last transmission: 7 December 1961
Total weight: 82 pounds
Diameter: 12 inches (S-15 gamma ray telescope)
6 inches (instrument column)
Height: 23.5 inches (S-15 gamma ray telescope)
20.5 inches (instrument column)
88 inches (total with fourth-stage Sergeant rocket)
Shell composition: Aluminum
Shape: Octagonal box

This successful first attempt at gamma ray astronomy eliminated the simultaneous matter-antimatter creation theory of steady-state cosmology.

Explorer S-45a
Launch vehicle: Juno 2
Launch and re-entry: 24 May 1961

Identical to the Explorer S-45, this Explorer did not achieve orbit due to launch vehicle failure.

Explorer S-55 (Meteoroid Satellite A)
Launch vehicle: Scout
Launch and re-entry: 30 June 1961
Total weight: 187 pounds
Diameter: 24 inches
Height: 76 inches

Wrapped around the Scout's nose, Explorer S-55 did not achieve orbit due to launch vehicle failure.

Explorer 12
Launch vehicle: Thor Delta
Launch: 16 August 1961
Re-entry: September 1963
Total weight: 83 pounds
Diameter: 26.2 inches
Height: 19 inches
Shell composition: Nylon honeycomb and fiberglass with an aluminum cover
Shape: Truncated cone attached to an octagonal platform

The spacecraft conducted particles and fields research, and its major accomplishment was identifying the Van Allen Belt as a magnetosphere.

Explorer 13 (Meteoroid Satellite B, S-55a)
Launch vehicle: Scout
Launch: 25 August 1961
Re-entry: 28 August 1961
Specifications: Same as Explorer S-55

The primary mission to test the Scout launch vehicle was successful, but the secondary micrometeoroid research mission was unsuccessful due to premature re-entry.

Explorer 14 (S-3A)
Launch vehicle: Thor Delta
Launch: 1 October 1962
Last transmission: 17 February 1964
Total weight: 89 pounds
Diameter: 27 inches
Height: 51 inches
Shell composition: Aluminum skin
Shape: Truncated cone mounted beneath an octagonal platform

A follow-on to Explorer 12, this spacecraft was designed to monitor corpuscular radiation, solar particles, cosmic radiation and solar winds and to correlate particle phenomena with magnetic-field observation. Launched into a highly elliptical orbit with an apogee of 61,000 miles and a perigee of 175 miles, the spacecraft returned excellent data for nearly a year.

Explorer 15 (S-3B)
Launch vehicle: Thor Delta
Launch: 27 October 1962
Last transmission: 19 May 1963
Specifications: Same as Explorer 14

Designed for the same role as Explorers 12 and 14, this spacecraft's despin mechanism failed, causing it to spin at 60 rpm rather than 10 rpm. The excessive spin rate made the directional detectors unusable and affected experiment results.

Explorer 16 (Meteoroid Satellite C, S-55)
Launch vehicle: 16 December 1962
Last transmission: 22 July 1963
Total weight: 222 pounds
Diameter: 24 inches
Height: 76 inches

With 160 beryllium-copper half-cylinder detectors covering an area of 30 square feet and of varying thicknesses (0.001 to 0.005 inch), this spacecraft yielded data on frequency of micrometeoroid penetration. The 0.001-inch sensors detected micrometeoroids 24 times, or 1.02 times per month, while 0.002-inch sensors were penetrated only 10 times, or 0.6 times per month.

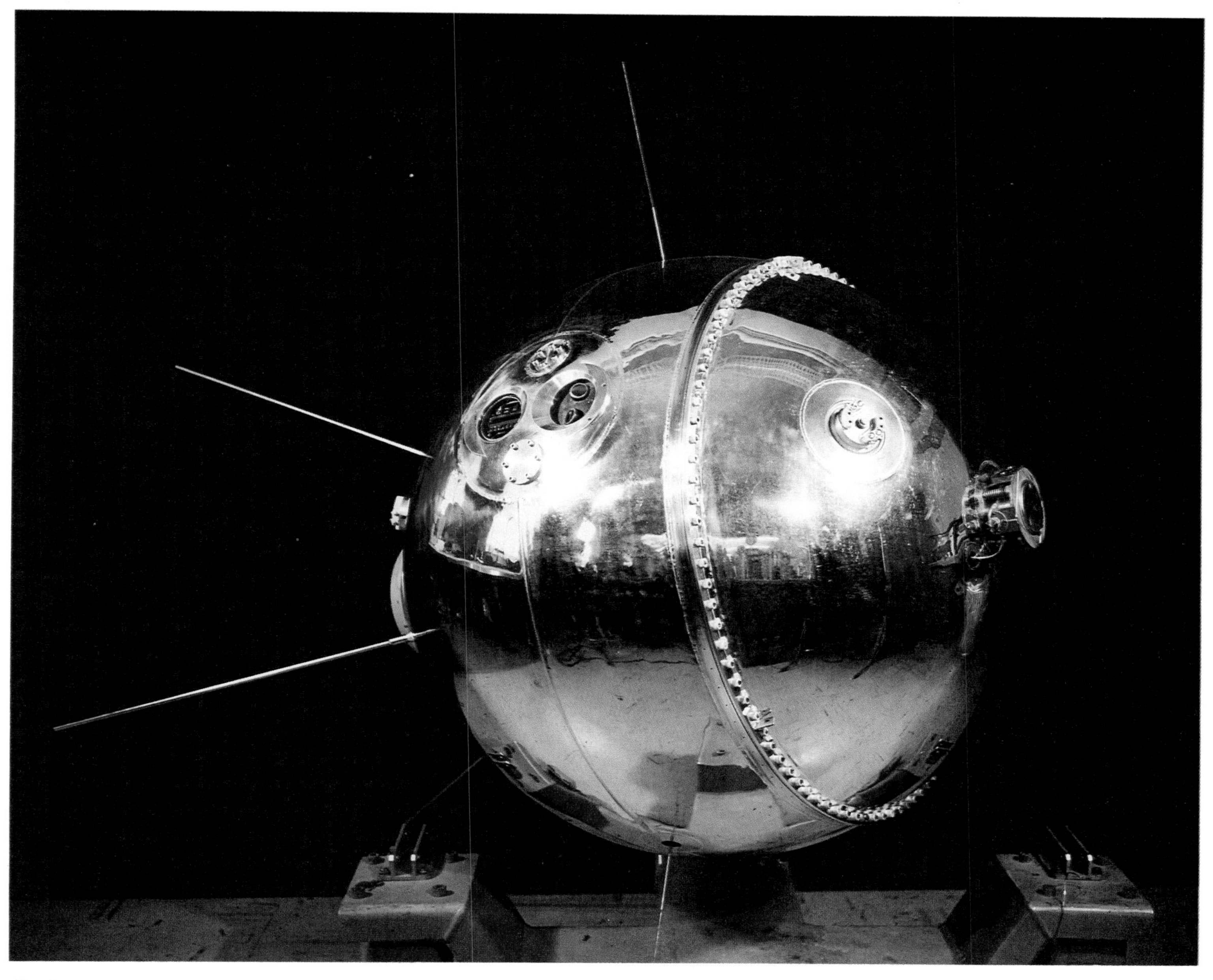

Atmospheric structure satellite Explorer 17.

Explorer 17
Launch vehicle: Thor Delta
Launch: 5 April 1963
Last transmission: 10 July 1963
Re-entry: 14 November 1966
Total weight: 405 pounds
Diameter: 35 inches
Shell composition: Stainless steel
Shape: Spherical

Designed to measure density, composition, pressure and temperature in the earth's atmosphere, this spacecraft discovered a belt of natural helium atoms around the earth.

Explorer S-66 (Beacon Explorer A)
Launch vehicle: Delta (DSV-3B)
Launch and re-entry: 19 March 1964
Total weight: 120 pounds
Diameter: 18 inches
Height: 12 inches
Shell composition: Honeycomb nylon and fiberglass

The Explorer S-66 was an octagonal spacecraft with a truncated pyramid covered with 360 fused silica reflectors. Planned optical tracking and geodesy with state-of-the-art lasers was not possible when orbit was not achieved due to launch vehicle failure.
(See also Explorers 22, 27)

Explorer 18
Launch vehicle: Delta (DSV-3C)
Launch: 26 November 1963
Last transmission: 12 May 1964
Re-entry: December 1965
Total weight: 138 pounds
Diameter: 28 inches
Height: 12 inches (two magnetometers on 7-foot booms, one on a 6-foot boom)

The first satellite launched from Cape Kennedy after the name was changed from Cape Canaveral, this spacecraft was similar in design and mission to Explorers 12, 14 and 15. It was the first of seven planned Interplanetary Monitoring Platform (IMP) satellites designed to observe and measure cosmic radiation, magnetic fields and solar wind in interplanetary space, data that was needed for the Apollo program. It discovered a region of high-energy radiation beyond the Van Allen Belt (see illustrations on page 48).

Explorer 19 (Air Density Explorer A)
Launch vehicle: Scout
Launch: 19 December 1963
Re-entry: 10 May 1981
Total weight: 18 pounds
Diameter: 12 feet
Shell composition: Aluminum foil and mylar
Shape: Spherical

Similar to the Explorer 9 balloon, this spacecraft was designed to measure upper-atmosphere air density in the polar

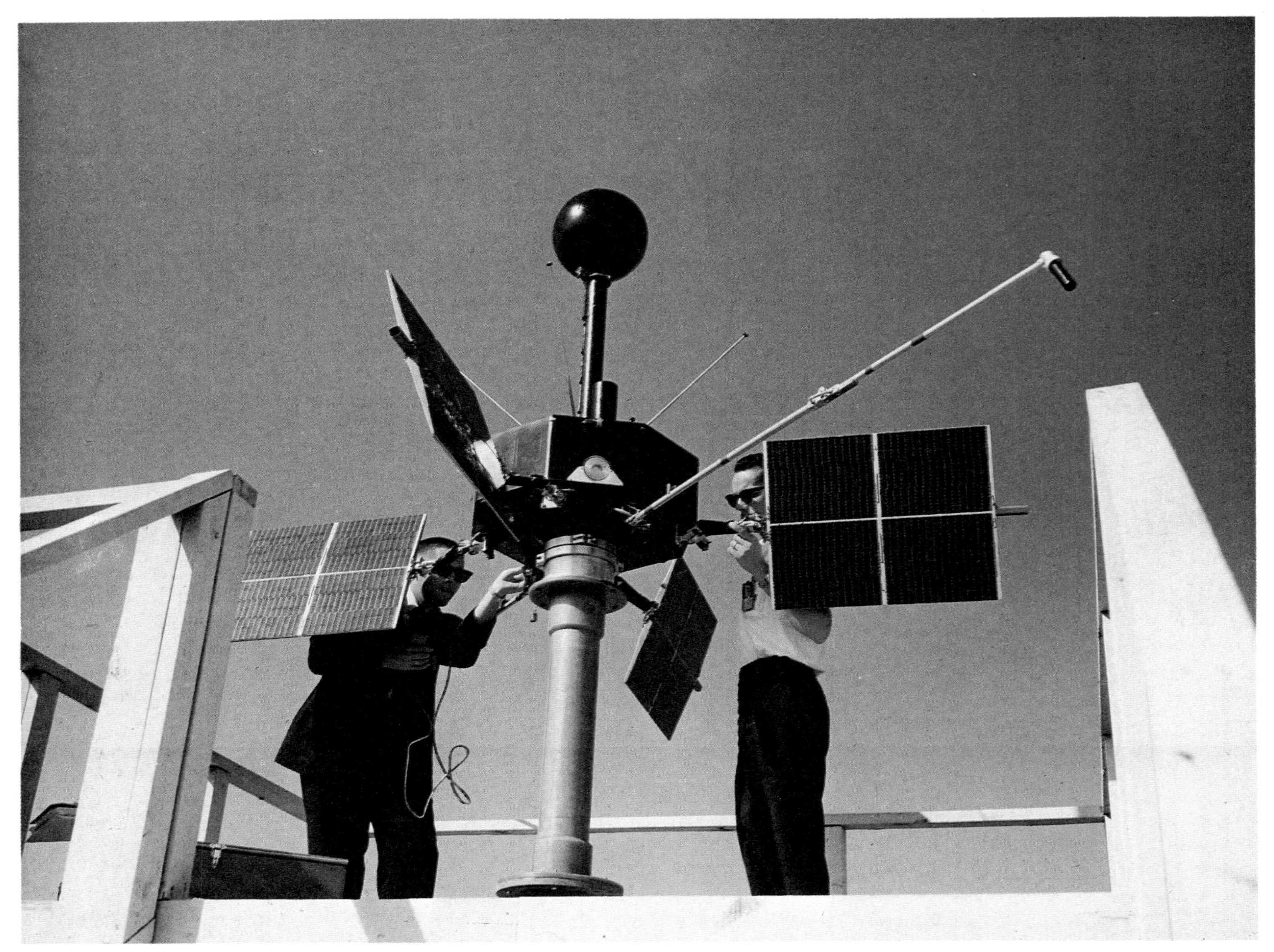

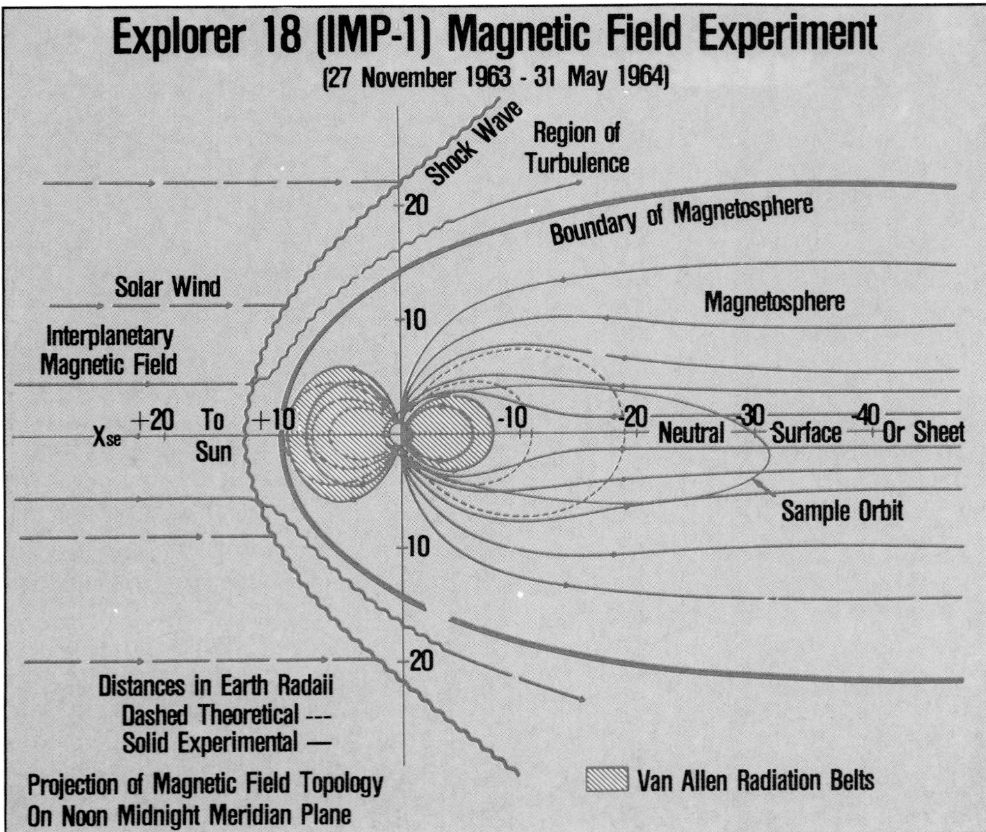

Above: Explorer 18 made observations over an extended period of the solar cycle.
Right: Explorer 20, with arms extended and experiment sphere atop the truncated cone.

regions. The tracking beacon failed, but the spacecraft was tracked optically.

Explorer 20 (Ionosphere Explorer)
Launch vehicle: Scout
Launch: 25 August 1964
Last transmission: 30 March 1966
Total weight: 97 pounds
Diameter: 26 inches
Height: 46.5 inches
Shell composition: Surface covered by solar cells
Shape: Cylindrical with truncated cone on bottom end and cone on top

Explorer 20 made radio soundings of the upper atmosphere and collected data on field-aligned ionization irregularities.

Explorer 21 (IMP-2)
Launch vehicle: Delta (DSV-3C)
Launch: 4 October 1964
Re-entry: 30 January 1966

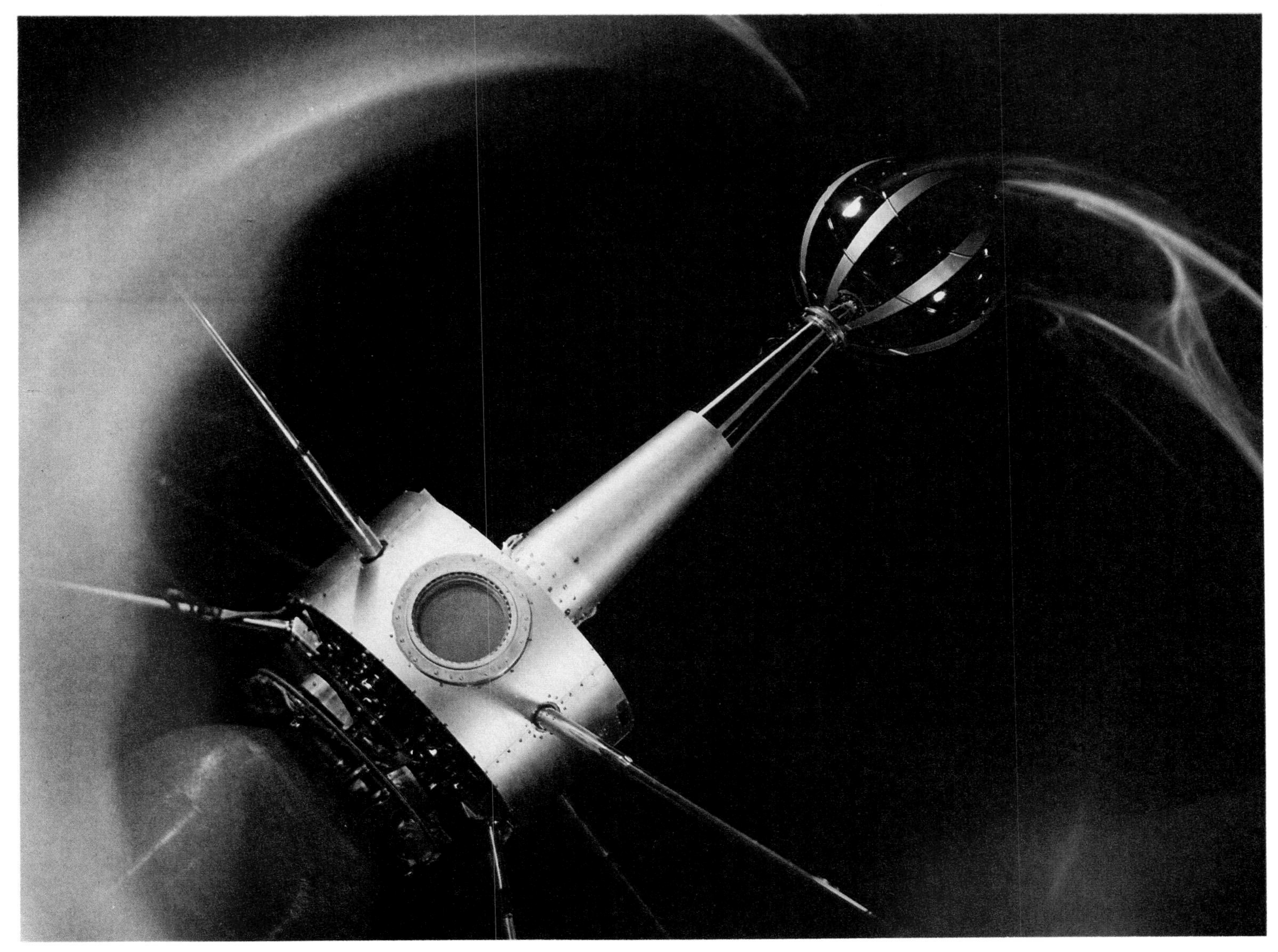

Total weight: 136 pounds
Diameter: 28 inches
Height: 8 inches
Shell composition: Solar panels
Shape: Octagonal

Explorer 21 conducted a detailed study of the cislunar space environment through cosmic rays, solar wind and magnetic field measurements.
(See also Explorers 18, 43)

Explorer 22 (Beacon Explorer B)
Launch vehicle: Scout
Launch: 10 October 1964
Total weight: 116 pounds
Diameter: 18 inches
Height: 12 inches
Shell composition: Honeycomb nylon and fiberglass
Shape: Octagonal main body with four solar panels extending from main body

Explorer 22 measured ionosphere structure and behavior through laser and doppler shift tracking experiments.
(See also Explorers S-66, 27)

Explorer 23 (Meteoroid Satellite D, S-55c)
Launch vehicle: Scout
Launch: 6 November 1964
Last transmission: 19 November 1964
Total weight: 295 pounds
Diameter: 24 inches
Height: 92 inches
Shape: Cylindrical

The Explorer 23 was a continuation of the S-55 series of experiments with stainless-steel micrometeoroid detectors.

Explorer 24 (Air Density Explorer B)
Launch vehicle: Scout
Launch: 21 November 1964
Last transmission: 25 July 1966
Re-entry: 18 October 1968
Total weight: 19 pounds
Diameter: 12 feet
Shell composition: Aluminum foil and mylar surface covered with 4000 white dots for passive thermal control
Shape: Spherical

Launched simultaneously with Explorer 25 in NASA's first dual-payload launch, this spacecraft was a continuation of the Explorer 9 and 19 missions. Explorer 24 returned data on radiation-air density relationships in the upper atmosphere of the North Polar region.
(See also Explorers 9, 19, 25, 39, 40)

Explorer 25 (Injun)
Launch data: Same as Explorer 24
Total weight: 90 pounds
Diameter: 24 inches
Shell composition: Aluminum shell with 40 sides, 30 covered with solar cells
Shape: Cylindrical tube

Launched at the same time as Explorer 24, this spacecraft also operated in conjunction with that spacecraft, gathering data on particle flux and energy in the same general area (the North Polar region) as the air density experiments were being conducted.
(See also Explorers 24, 39, 40)

Explorer 26
Launch vehicle: Delta (DSV-3C)
Launch: 21 December 1964

The Explorer Program at a Glance

Explorer Number	Scientific Experiment	Prime Contractor(s)	Launch Site
1958			
1	Particles and fields	US Army	CC
2	Particles and fields	US Army	CC
3	Particles and fields	US Army	CC
4	Radiation mapping	US Army	CC
5	Particles and fields	US Army	CC
1959			
S-1	Particles and fields	NASA/US Army	CC
6	Particles and meteorology	NASA/US Army	CC
7	Energetic particles	NASA/US Army	CC
1960			
S-46	Energetic particles	NASA/US Army	CC
8	Ionosphere	NASA/US Army	CC
S-56	Atmospheric physics	NASA/US Army	WI
1961			
9	Atmospheric physics	NASA/US Army	WI
S-45	Ionosphere	NASA/US Army	CC
10	Particles and fields	NASA	CC
11	Gamma ray astronomy	NASA/US Army	CC
S-45a	Ionosphere	NASA/US Army	CC
S-55	Meteoroids	NASA	WI
12	Particles and fields	NASA	CC
13	Meteoroids	NASA	WI
1962			
14	Particles and fields	NASA	CC
15	Particles and fields	NASA/AT&T	CC
16	Micrometeoroids	NASA	WI
1963			
17	Aeronomy	NASA	CC
18	Particles and fields	NASA	CK
19	Atmospheric physics	NASA	VAFB
1964			
20	Ionosphere	NASA	VAFB
21	Particles and fields	NASA	VAFB
22	Ionosphere	APL	VAFB
23	Micrometeoroids	NASA	WI
24	Atmospheric physics	NASA	VAFB
25	Atmospheric physics	U of Iowa	VAFB
26	Particles and fields	NASA	CK
1965			
27	Geodesy	APL	WI
28	Particles and fields	NASA	CK
29	Geodesy	APL	CK
30	Solar physics	NRL	WI
31	Ionosphere	NRL	VAFB
1966			
32	Aeronomy	NASA	CK
33	Particles and fields	NASA	CK
1967			
34	Particles and fields	NASA	VAFB
35	Particles and fields	NASA	CK
1968			
36	Geodesy	APL	VAFB
37	Solar X-ray emissions	NRL	WI
38	Radio astronomy	Fairchild Hiller	VAFB
39	Atmospheric physics	NASA	VAFB
40	Atmospheric physics	U of Iowa	VAFB
1969			
41	Particles and fields	NASA	VAFB
1970			
42	Celestial X-ray	APL	SM
1971			
43	Particles and fields	NASA	CK
44	Solar emissions	NRL	WI
45	Magnetic storms	NASA	SM
1972			
46	Meteoroids	LTV	WI
47	Particles and fields	NASA	KSC
48	Gamma rays	NASA	SM
1973			
49	Astronomy	NASA	KSC
50	Particles and fields	EMR	KSC
51	Atmospheric physics	RCA	VAFB
1974			
52	Magnetosphere	U of Iowa	VAFB
1975			
53	Astronomy	NASA	SM
54	Atmospheric physics	RCA	VAFB
55	Atmospheric physics	RCA	KSC

Last transmission: 21 January 1967
Total weight: 101 pounds
Diameter: 27.75 inches
Height: 17 inches
Shape: Octagonal platform atop a truncated cone

Four solar panels extended from the sides and a 34-inch tube supporting a magnetometer was mounted atop the spacecraft. Explorer 26 operated five experiments identical to those of Explorer 25, studying particle activity in the Van Allen Belt.

Explorer 27 (Beacon Explorer C)
Launch vehicle: Scout
Launch: 29 April 1965
Total weight: 132 pounds
Diameter: 48 inches
Height: 12 inches
Shell composition: Honeycomb nylon and fiberglass
Shape: Octagonal

This spacecraft continued the ionosphere measurements of Explorer 22

(Beacon Explorer B), studying gravimetric geodesy and measuring electron density and temperature through laser tracking.

Explorer 28 (IMP-C)
Launch vehicle: Delta (DSV-3C)
Launch: 29 May 1965
Re-entry: 4 July 1968
Total weight: 130 pounds
Diameter: 28 inches
Height: 8 inches
Shape: Octagonal main structure

Explorer 28 consisted of a magnetometer sphere mounted on a telescopic 6-foot boom on top of the spacecraft and two magnetometers on 7-foot booms that extended from the body, along with four solar panels and four telemetry antennae. This spacecraft continued the IMP study of solar-terrestrial relationships, especially magnetosphere bounding and cislunar radiation environment.
(See also Explorers 18, 21)

Explorer 29 (GEOS-A)
Launch vehicle: Thrust Augmented Delta (TAD)
Launch: 6 November 1965
Last transmission: 16 January 1967
Total weight: 385 pounds
Diameter: 48 inches
Height: 32 inches
Shell composition: Aluminum

This spacecraft, also known as Geodetic Explorer A, was part of the US Geodetic Satellite Program, a joint effort of NASA, DOD and the US Department of Commerce. It was an octagonal shell with an eight-sided truncated pyramid on top. An extendable 60-foot gravity gradient boom of silver-plated beryllium copper was mounted on top, with an eddy current damper on the end of the boom. A 24-inch fiberglass hemisphere with a broad-band spiral antenna paint pattern projected from the satellite's base. Xenon flash tubes, quartz-cube corner reflectors and 4-inch conical antennae also were mounted on the base. The craft was designed to integrate geodetic data into a single one-world system, relating them all to the earth's center of mass so that site locations could be determined in a three-dimensional coordinate system accurate to 33 feet or better. Its function was also to map the earth's irregular gravitational field and to operate and correlate data from five geodetic measuring systems.

Explorer 30 (Solar Explorer A)
Launch vehicle: Scout
Launch: 19 November 1965
Last transmission: 7 November 1967
Total weight: 125 pounds
Diameter: 24 inches
Shape: Two 24-inch hemispheres separated by a 3.5-inch equatorial band

Explorer 30 monitored solar X-rays and ultraviolet emissions during 1964-65 International Quiet Sun Years (IQSY) under the directions of the Naval Research Laboratory (NRL).

Explorer 31
Launch vehicle: Thor Agena B
Launch: 29 November 1965 (in conjunction with the Canadian Alouette 2)
Total weight: 218 pounds
Diameter: 30 inches
Height: 25 inches
Shell composition: 15 percent of surface covered by solar cells
Shape: Octagonal

Explorer 31 worked with Alouette 2 to make related studies of the ionosphere in areas not yet explored.

Explorer 32 (Atmosphere Explorer B)
Launch vehicle: Delta (DSV-3C-1A)
Launch: 25 May 1966
Last transmission: 31 March 1967
Total weight: 495 pounds
Diameter: 35 inches
Shell composition: 0.025-inch stainless steel
Shape: Spherical

A follow-on to Explorer 27 (Atmosphere Explorer A), Explorer 32 studied helium and hydrogen in the upper atmosphere with longer-life solar cells.

Explorer 33 (IMP-D)
Launch vehicle: Thor Delta (TAD)
Launch: 1 July 1966
Total weight: 206 pounds
Diameter: 28 inches
Height: 8 inches (with 6-foot magnetometer)
Shape: Octagonal

This spacecraft, failing in its attempt to become the first IMP to achieve lunar orbit, went into earth orbit and returned considerable data on interplanetary space to supplement the data returned by Explorers 18, 21 and 28.

Explorer 34 (IMP-F)
Launch vehicle: Thrust Augmented Delta (TAD)
Launch: 24 May 1967
Re-entry: 3 May 1969
Total weight: 163 pounds
Diameter: 28 inches
Height: 10 inches

Explorer 34 was an improved version of Explorers 18, 21 and 28. The main body was an octagon. Two 6-foot magnetometer booms and four solar panels extended from the midsection and four telemetry antennae projected from the top. Similar to earlier IMP spacecraft, it carried out 11 experiments to measure solar and galactic cosmic rays in the magnetosphere and interplanetary space.

Explorer 35 (IMP-E)
Launch vehicle: Delta
Launch: 19 July 1967
Total weight: 230 pounds (including 80-pound retromotor)
Diameter: 28 inches (main body)
Height: 8 inches
Shape: Octagonal

Explorer 35 is the same as Explorer 33 and similar to Explorers 18, 21 and 28, except that the retrorocket was mounted atop the spacecraft in place of the usual magnetometer boom. Four whip antennae projected from the top and two 6-foot magnetometer booms and four solar panels extended from the main body. This was the first of the IMP series interplanetary Explorers to orbit the moon, achieving lunar orbit on 22 July 1967. It returned data on the earth's magnetosphere and observed no corresponding lunar magnetic field or radiation belt.

Explorer 36 (GEOS-B)
Launch vehicle: Delta (DSV-3E)
Launch: 11 January 1968
Total weight: 460 pounds
Diameter: 48 inches
Height: 32 inches
Shell composition: Aluminum
Shape: Octagonal

Explorer 36 was a continuation of the US Geodetic Satellite Program activities started by the nearly identical Explorer 29 (GEOS-A).
(See also Explorer 29)

Explorer 37 (Solar Explorer B)
Launch vehicle: Scout
Launch: 5 March 1968
Last transmission: 16 March 1970
Total weight: 198 pounds
Diameter: 30 inches
Height: 27 inches
Shape: Cylindrical

Explorer 37 consisted of a cylinder mounted on a central band with X-ray photometers, Geiger tubes, photomultipliers, a solar aspect system, and altitude control and spin nozzles around the middle. Solar cells, contained in 24 panel arrays 7 inches by 10 inches, covered the remainder of the vertical sides. Four radio antennae extended from the top, and four from the bottom. This second joint NASA/NRL spacecraft was used to measure and monitor selected solar X-ray and ultraviolet emissions.
(See also Explorer 30)

Explorer 38 (Radio Astronomy Explorer A)
Launch vehicle: Thrust Augmented Improved Delta (TAID)
Launch: 4 July 1968
Full antenna deployment: 8 October 1968
Total weight: 417 pounds

Explorer 38 was a cylindrical aluminum structure with four 5-foot-long helical solar panels deployed around the sides and an apogee motor nozzle extended from the aft end. Sensors and telemetry antennae were mounted on the outer edges of the solar panels. Appendages (extended when the craft reached final orbit) included two 0.75-pound yo-yo weights attached to 27-foot wires for despinning the spacecraft from an initial 92 rpm to 2 rpm before being cut loose (the final despinning was accomplished by three electromagnetic coils); a 630-foot-long (maximum) vibration damper extending from the forward end (toward earth); a 120-foot-long dipole antenna extended in two 60-foot sections from the sides along earth horizontal; four 750-foot-long antennae arranged in two 60-degree angles, one extending toward the earth, the other toward space. Each antenna boom was stored as a 0.002-inch thick beryllium-copper alloy tape, deployed from motor-driven reels to form 0.5-inch-diameter tubes. The gravity gradient was stabilized.

The mission of this first Explorer of its type was to monitor low-frequency radio signals from cosmic sources from our solar system (particularly strong radio bursts from the region of Jupiter) and from the earth's magnetosphere. Instrumentation included three nine-step radiometers measuring frequencies between 0.45 and 9.2 MHz and two burst receivers measuring the 0.2 to 5.53 MHz range and 0.245 to 3.93 MHz range, respectively.
(See also Explorer 49)

Explorer 38 is mated to the TAID third stage.

Explorer 39 (Air Density Explorer C)
Launch vehicle: Scout
Launch: 8 August 1968
Re-entry: 22 June 1981
Total weight: 20.8 pounds

Explorer 39 consisted of a 12-foot polka-dotted sphere made of four alternating layers of 0.0005-inch-thick aluminum foil and 0.0005-inch-thick mylar, with the foil comprising the outer layer. The sphere was constructed by bonding together 40 flat triangular sections. It was divided at the equator by a strip of plastic, allowing both metallic hemispheres to serve as antennae for the tracking beacon. It was inflated in orbit with nitrogen gas, which was then permitted to escape, allowing the Explorer 39 spacecraft to maintain its shape using its own rigidity.

This Air Density Explorer was launched in conjunction with Injun 5 (Explorer 40) in the same way that Air Density Explorer B (Explorer 24) had been launched in conjunction with Injun 4 (Explorer 25). The two spacecraft carried out coordinated radiation studies of the upper atmosphere.
(See also Explorers 24, 25, 40)

Explorer 40 (Injun 5)
Launch vehicle: Scout
Launch: 8 August 1968
Total weight: 157 pounds
Diameter: 30 inches
Height: 29 inches
Shell composition: Outside surface covered by solar cells
Shape: Hexagonal cylinder

Mounted inside and extending slightly beyond the forward end was a cylindrical tube in which Explorer 39 was carried folded into orbit, together with ejection and inflation equipment. Appendages included five hinged booms, carried folded during launch. Explorer 40 completed successful upper-atmosphere radiation studies in conjunction with Explorer 39.
(See also Explorers 24, 25, 39)

Explorer 41 (IMP-G)
Launch vehicle: Thrust Augmented Improved Delta (TAID)
Launch: 21 June 1969
Re-entry: 23 December 1972
Total weight: 174 pounds
Diameter: 28 inches (with four 16-inch transmitting antennae and two hinged 6-foot magnetometer booms)
Shape: Octagonal

Explorer 41 continued earlier IMP missions.
(See also Explorers 18, 21, 28, 33, 34, 35)

Explorer 42 (SAS-A)
Launch vehicle: Scout
Launch: 12 December 1970
Re-entry: 5 April 1979
Total weight: 315 pounds
Diameter: 2 feet
Height: 2 feet
Shape: Cylindrical drum

Explorer 42 was launched jointly by NASA and the University of Rome from the Italian San Marco launch platform in the Indian Ocean off Kenya, the first US spacecraft launched by a foreign country. A Small Astronomy Satellite (SAS), its mission was to catalog celestial X-ray sources within and beyond the Milky Way to correlate with the findings of radio and optical astronomy.

Explorer 43 (IMP-I)
Launch vehicle: Delta M-6
Launch: 13 March 1971
Re-entry: 2 October 1974
Total weight: 635 pounds (including 215 pounds of experiments)
Diameter: 4 feet 6 inches
Height: 6 feet
Shape: 16-sided drum

Explorer 43 provided a continuation of earlier IMP missions.
(See also Explorers 18, 21, 28, 33, 34, 35, 41)

Explorer 44 (SOLRAD 10)
Launch vehicle: Scout
Launch: 8 July 1971
Re-entry: 15 December 1979
Total weight: 260 pounds
Diameter: 30 inches
Height: 23 inches
Shape: 12-sided

A joint NASA/NRL project, SOLRAD was designed to monitor solar X-ray and ultraviolet emissions to better understand solar processes and improve the prediction of solar activity and ionospheric disturbances.

Explorer 45 (SSS-A)
Launch vehicle: Scout
Launch: 15 November 1971
Total weight: 114 pounds
Diameter: 27 inches (with one 30-inch, two 24-inch and two 108-inch booms, and four 24-inch antennae)
Shape: Octagonal

The role of this second Italian-launched US spacecraft was to measure aerodynamic heating and radiation damage

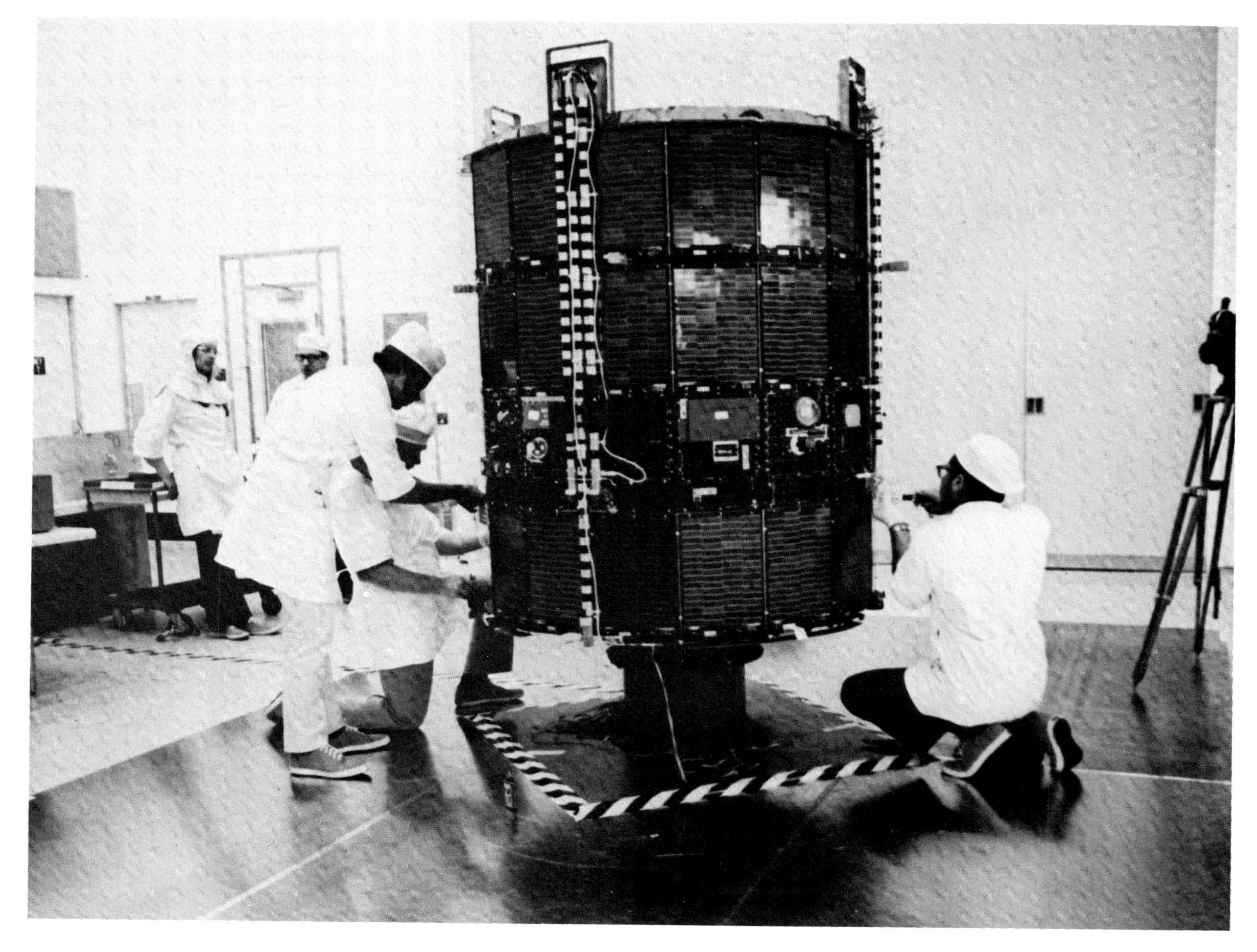

Atmosphere Explorer 51 studied the thermosphere.

during launch, and to investigate particles and magnetic and electric fields in the magnetosphere.

Explorer 46
Launch vehicle: Scout
Launch: 13 August 1972
Re-entry: 2 November 1979
Total weight: 300 pounds

Explorer 46 was a windmill-like spacecraft with meteoroid bumper panels deployed from a hexi-cylindrical bus; the fourth stage of the booster also orbited while attached to the spacecraft. Panels were 10 feet 6 inches long.

A Meteoroid Technology Satellite similar to Explorers S-55, 13, 16 and 23, this spacecraft carried 96 stainless-steel detectors and 64 capacitor detectors. About 20 impacts were recorded through December 1972.
(See also Explorers S-55, 13, 16, 23)

Explorer 47 (IMP-H)
Launch vehicle: Delta 1640
Launch: 22 September 1972
Total weight: 860 pounds
Description: 16-sided drum with boom-mounted experiment packages

Explorer 47 continued earlier IMP missions.
(See also Explorers 18, 21, 28, 33, 34, 35, 41, 43)

Explorer 48 (SAS-B)
Launch vehicle: Scout
Launch: 16 November 1972
Re-entry: 20 August 1980
Total weight: 410 pounds
Height: 51 inches
Width: 156 inches (from tip to tip of solar panels)
Shape: Cylindrical

The mission of this SAS was to study gamma rays with a 32-level digitized spark-chamber telescope mounted in a dome. This gamma ray telescope had a sensitivity 10 times greater than any other gamma ray detector ever placed in orbit.

Explorer 49 (Radio Astronomy Explorer B)
Launch vehicle: Delta
Launch: 10 June 1973
Total weight: 723 pounds at launch
442 pounds in orbit
Diameter: 36 inches
Height: 31 inches
Shape: Cylindrical

Placed in lunar orbit on 15 June 1973, this spacecraft conducted investigations of low-frequency signals from galactic and extragalactic sources and from the sun, Earth and Jupiter. Explorer 49 monitored the 0.02 to 13 MHz range from a vantage point where extraneous radio activity from earth caused a minimum of interference.

Explorer 50 (IMP-J)
Launch vehicle: Delta
Launch: 25 October 1973

Total weight: 818 pounds
Diameter: 53 inches
Height: 62 inches
Shape: 16-sided drum

This was the last of the IMP spacecraft that had begun operation in 1963 with Explorer 18. IMP-J carried 12 experiments similar to those of the previous IMP spacecraft as well as an experimental solar panel with COMSAT labs violet solar cells, and a new type of experimental data multiflex unit.

Explorer 47 conducted its studies from an orbit approximately half way to the moon.

(See also Explorers 18, 21, 28, 33, 34, 35, 41, 43, 47)

Explorer 51 (Atmosphere Explorer C)
Launch vehicle: Delta
Launch: 16 December 1973
Re-entry: 12 December 1978
Total weight: 1450 pounds
Diameter: 53 inches
Height: 45 inches
Shape: Cylindrical

The mission of Explorer 51 was to investigate photochemical processes accompanying the absorption of solar ultraviolet radiation in the earth's atmosphere by making closely coordinated measurements of reacting constituents with an onboard propulsion system permitting change of spacecraft apogee and perigee on command.
(See also Explorers 27, 32, 54, 55)

Explorer 52 (Hawkeye)
Launch vehicle: Scout

Atmosphere Explorer 52 (*Hawkeye*) carried 14 scientific experiments.

Launch: 3 June 1974
Total weight: 58.7 pounds
Diameter: 30 inches at base, 10 inches at top
Height: 30 inches

An eight-sided truncated cone, the spacecraft structure provided mounting surfaces to support electronic and mechanical packages for experiments. Solar-cell panels were mounted on all sides and on the bottom. Several sensors and booms projected from the spacecraft to support experiments: two electric field antenna booms extended to 75 feet and a fluxgate magnetometer boom and a search coil magnetic antenna were extendable to 5 feet.

Hawkeye's primary objective was to investigate the interaction of solar wind on the earth's magnetic field, with emphasis on the North Polar region.

Explorer 53 (SAS-C)
Launch vehicle: Scout
Launch: 7 May 1975
Re-entry: 9 April 1979
Total weight: 430 pounds
Diameter: 26 inches
Height: 24 inches

This spin-stabilized satellite carried its scientific payload in a separately fabricated compartment fitted to a common bus section. Bus was standard for all missions, and contained mission-support subsystems such as power, attitude control, communications and data storage. Four foldable solar panels provided power. Stability was provided by a reaction wheel and a passive nutation damper. Sun sensors, earth horizon sensors and a star sensor, as well as a magnetic torque control system could point the spacecraft in any direction in the sky or vary its spin rate on command.

The mission of this third SAS was to study X-ray sources within and beyond the Milky Way galaxy as part of an investigation of high-energy astrophysics.

Explorer 54 (Atmosphere Explorer D)
Launch vehicle: Delta 2910
Launch: 6 October 1975
Re-entry: 12 March 1976
Total weight: 1488 pounds
Diameter: 53.2 inches
Height: 45 inches

A drum-shaped, 16-sided polyhedron, Explorer 54 consisted of two shells, inner and outer, with solar cells, telemetry antennae and viewing ports on the outer shell. The inner shell held 12 scientific instruments and four engineering measurements (212 pounds); electronic packages; an attitude-control system and hydrazine thruster subsystems. Solar cells on the exterior shell and redundant nickel-cadmium batteries provided 120 watts of power. The craft continued the work of Explorer 51.
(See also Explorers 27, 32, 51, 55)

Explorer 55 (Atmosphere Explorer E)
Launch vehicle: Delta 2910
Launch: 20 November 1975
Re-entry: 18 June 1981
Total weight: 1587 pounds
Specifications and description: Same as Explorer 54

The final spacecraft in the numbered Explorer series that spanned nearly two decades, this satellite was also the last in a subseries of three atmosphere Explorers that had begun with Explorer 51 (Atmosphere Explorer C). The Explorer 55 mission was generally the same as Atmosphere Explorers C and D, with the added task of measuring ozone in the upper atmosphere and its possible depletion from man-made causes.
(See also Explorers 27, 32, 51, 54; Applications Explorer Mission B)

Explorer, Active Magnetosphere Particle Tracer

(See AMPTE)

Explorer, Applications

(See Applications Explorer Missions)

Explorer, Dynamics

(See Dynamics Explorer)

Explorer S

(See Explorer, general introductory remarks)

Ferret

Launch vehicle: Thor Agena
Estimated weight: 500 pounds

This Lockheed/Sanders-developed spacecraft series (Code 711) is composed of classified second-generation US Air

Galaxy is based on Hughes' HS 376 spacecraft design and has 24 transponders.

Force electromagnetic reconnaissance satellites that will be superseded by a third-generation Hughes-designed satellite in the late 1980s.
(See also ELINT, Rhyolite)

Fire

Launch vehicle: Atlas X259
Launch: 14 April 1964 (Fire 1)
22 May 1965 (Fire 2)
Re-entry: 14 April 1964 (Fire 1)
22 May 1965 (Fire 2)

These were a pair of re-entry tests designed to investigate the heating environment encountered by a spacecraft entering the earth's atmosphere at high speed. Actual speeds were 25,884 mph (Fire 1) and 25,399 mph (Fire 2), with excellent data obtained.

FLTSATCOM

The US Navy Fleet Satellite Communications (FLTSATCOM) system was designed to provide worldwide high-priority UHF communications between aircraft, ships, submarines and ground stations, as well as between the US military services and the presidential command network. The communications subsystem provided more than 30 voice and 12 teletype channels designed to serve small mobile users as well as major centers. The payload of the spacecraft included UHF and SHF communications equipment and antennae. The US Air Force AFSATCOM system was carried aboard these Navy spacecraft, built by the TRW Defense and Space Systems Group.
(See also AFTSATCOM)

FLTSATCOM (General Specifications)
Launch vehicle: Atlas Centaur
Launches: 9 February 1973 (FLTSATCOM A)
4 May 1979 (FLTSATCOM B)
17 January 1980 (FLTSATCOM C)
30 October 1980 (FLTSATCOM D)
6 August 1981 (FLTSATCOM E)
Total weight: 4136 pounds at launch
2216 pounds in orbit
Diameter: 8 feet (30 feet with solar arrays fully deployed)
Height: 4 feet 2 inches (16 feet with antennae)
Shape: Hexagonal
Systems: One 25 KHz (SHF) fleet broadcast channel
Nine 25 KHz (UHF) fleet relay channels
Twelve 5 KHz (UHF) USAF narrow-band channels
One 500 KHz (UHF) DOD wide-band channel
Receiving band: 292-400 MHz (240-400 MHz from FLTSATCOM C on)
Transmission band: 244-279 MHz (240-400 MHz from FLTSATCOM C on)

The basic FLTSATCOM consisted of two stacked hexagonal modules, antennae, and two winglike solar arrays. It had an 18-turn helical UHF receive antenna and a 15-foot deployable parabolic UHF transmit antenna. The solar array provided 1200 watts of power. Axis stabilization was provided by redundant, body-mounted momentum wheels that interacted with monopropellant hydrazine thrusters to stabilize the spacecraft's attitude and point the antennae at the earth's center.

Fordsat

Launch vehicle: Space shuttle
First launch: 1988
Total weight: 2450 pounds

Developed by Ford Aerospace, these spacecraft are described as fixed service C/Ku band communications satellites.

Galaxy

Launch vehicle: Delta
Launch: 28 June 1983 (Galaxy 1)
22 September 1983 (Galaxy 2)
21 September 1984 (Galaxy 3)
Total weight: 1222 pounds
Number of transponders: 24
Frequency: 4-6 GHz
Signal strength: 34 dBW
Expected lifetime: Nine years

This series of communications were built by Hughes Aircraft and launched by

NASA from Cape Canaveral on a reimbursable basis. The function of the Galaxy spacecraft, which are owned by Hughes Communications, is to provide domestic television communications.
(See HS 376)

Galileo

Launch vehicle: Space shuttle
Launch: 1986
Total weight: 5500 pounds

Developed by NASA's Jet Propulsion Laboratory, the Galileo spacecraft was designed to be placed into orbit around the planet Jupiter. The spacecraft will be a follow-on to earlier Pioneer and Voyager spacecraft that conducted scientific surveys of Jupiter.

Gemini

The Gemini program was America's second manned spacecraft program designed to bridge the technological gulf between the early Mercury program and the Apollo lunar-landing program. Like the Mercury spacecraft, the Gemini spacecraft were produced by McDonnell Douglas. Where the Mercury spacecraft accommodated only one crew member, the Gemini spacecraft carried two. The cabin area of Gemini was 50 percent greater than that of Mercury. Where Mercury had been a research vehicle flown for the most part by automatic ground control with only secondary manual override from the astronaut, Gemini was a spacecraft designed to be flown primarily by its crew. An important part of the Gemini mission was training astronauts to maneuver in space and to rendezvous and dock one spacecraft with another. All this was in anticipation of President John Kennedy's announced goal of putting an American on the moon by the end of the 1960s.

McDonnell had the first Gemini mockup available for NASA inspection on 29 March 1962 and on 2 April 1963, NASA officially ordered 13 flight-rated Gemini spacecraft, two mission simulator trainers and eight nonflying spacecraft for ground tests. The first of the flight-rated spacecraft (Gemini 1) was delivered to Cape Canaveral by McDonnell on 4 October 1963 and the first simulator was delivered on 13 December. The first unmanned Gemini test was successfully conducted on 8 April 1964. Much of the program consisted of exercises in which the Gemini spacecraft were manually maneuvered to

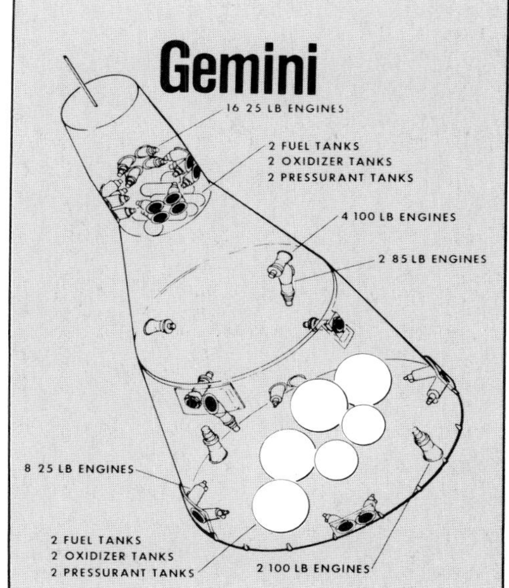

Above: **Onboard propulsion systems give Gemini its versatile flight response.**
Far left: **Gemini 9A was launched on 1 June 1966.**
Left: **Close-up view of the space sextant used for the Gemini 7 Simple Navigation (D-9) Experiment.**
Next page: **Gemini 6A rendezvous with Gemini 7.**

rendezvous with another Gemini (Gemini 6A and 7) or rendezvous and dock with a target vehicle. The target vehicle used (from Gemini 8 on) was an Atlas Agena upper stage, a 1700-pound unit known as the Augmented Target Docking Adapter (ATDA). The EVA in which astronauts maneuvered in space outside the spacecraft was conducted on most of the Gemini flights.

Height: (Manned re-entry module) 18.4 feet
(Aft adapter module) 19 feet
Diameter: (Adapter module base) 10 feet
(Adapter re-entry module junction) 90 inches
(Top of re-entry module) 39 inches
Total weight: (Re-entry module) variable: 7000 to 8374 pounds
(Adapter module) 4400 pounds
Material: (Re-entry module) Titanium structure with forward section skin of beryllium and cabin section hatch covered with nickel alloy
(Adapter module) Frame of magnesium, aluminum and titanium with magnesium skin

Gemini 1
Launch vehicle: Titan 2
Launch: 8 April 1964
Re-entry: 12 April 1964
Crew: None

The first launch of the Gemini program was primarily for the purpose of testing the launch vehicle and its compatibility with the spacecraft. The spacecraft was not equipped to separate from the second stage and was not recovered.

Gemini 2
Launch vehicle: Titan 2
Launch and recovery: 19 January 1965
Crew: None

The second Gemini flight test demonstrated the structural integrity of the systems and performance of the spacecraft throughout flight, re-entry and parachute water landing.

Gemini 3
Launch vehicle: Titan 2
Launch and recovery: 23 March 1965
Crew: Virgil Grissom, John Young

The first manned Gemini flight completed three orbits in 4 hours 53 minutes. Even though the flight was shorter than either of the last two Mercury flights, it was a milestone in that it was the first time a crew had manually controlled a maneuver in space and conducted a manually controlled re-entry.

Gemini 4
Launch vehicle: Titan 2
Launch: 3 June 1965
Recovery: 7 June 1965
Crew: James McDivitt, Edward White

The second of the manned Gemini flights completed 62 orbits in 97 hours 56 minutes, with astronaut White spending 36 minutes outside the spacecraft in the historic first American EVA in space. Eleven scientific experiments were successfully completed, but a practice rendezvous with the booster was not.

Gemini 5
Launch vehicle: Titan 2
Launch: 21 August 1965
Recovery: 29 August 1965
Crew: Charles Conrad, Gordon Cooper

This mission completed 120 orbits in 190 hours 56 minutes. There were 17 experiments scheduled – 6 scientific, 6 DOD, 5 medical and one engineering, of which 16 were successfully completed.

Gemini 6
Launch vehicle: Atlas Agena
Launch: 25 October 1965
Crew: None

This unmanned spacecraft, designed to test rendezvous and docking capability, was not placed into orbit because the launch vehicle exploded on ignition.

Gemini 6A
Launch vehicle: Titan 2
Launch: 15 December 1965
Recovery: 16 December 1965
Crew: Walter Schirra, Thomas Stafford

This spacecraft completed 15 orbits in 25 hours 51 minutes, and while in space successfully maneuvered to a rendezvous with Gemini 7, which had been launched

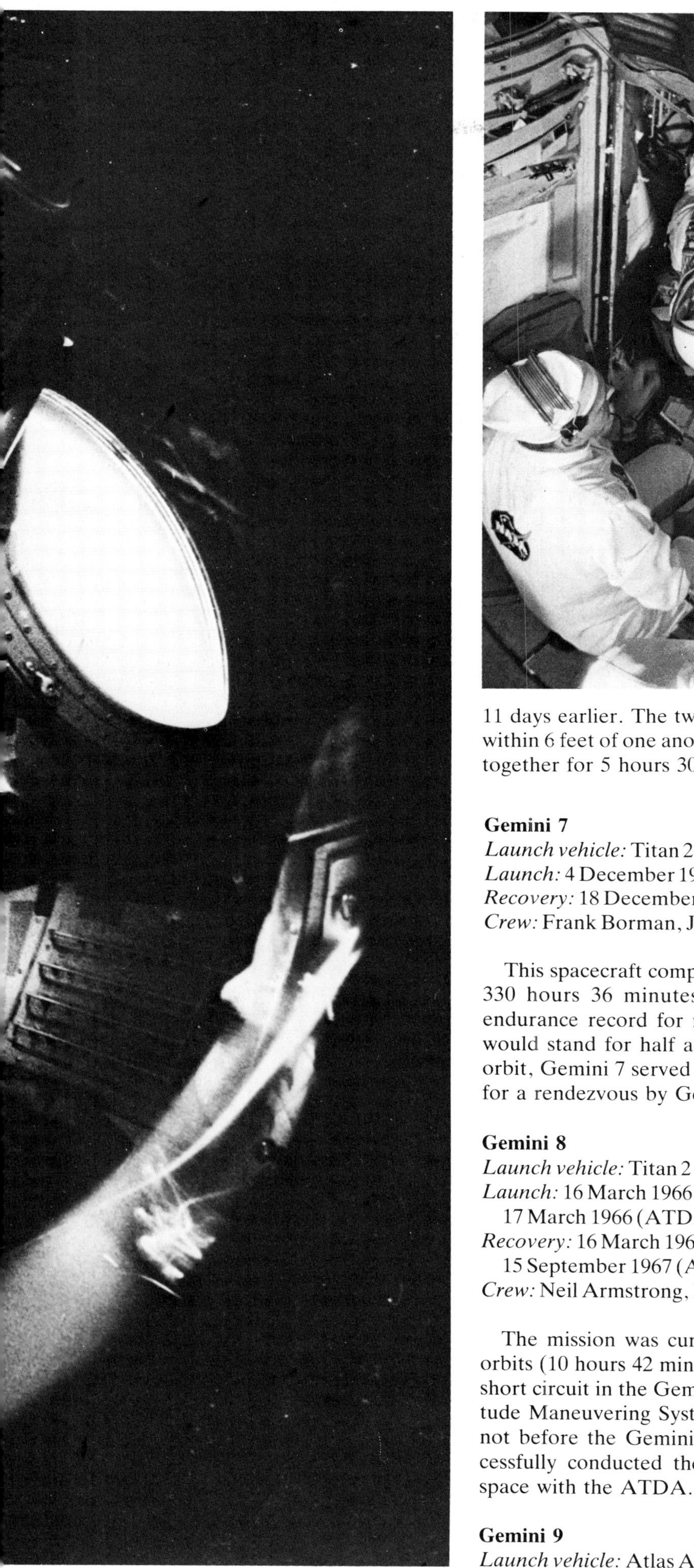

11 days earlier. The two spacecraft came within 6 feet of one another and remained together for 5 hours 30 minutes.

Gemini 7
Launch vehicle: Titan 2
Launch: 4 December 1965
Recovery: 18 December 1965
Crew: Frank Borman, James Lovell

This spacecraft completed 206 orbits in 330 hours 36 minutes, establishing an endurance record for man in space that would stand for half a decade. While in orbit, Gemini 7 served as a passive target for a rendezvous by Gemini 6A.

Gemini 8
Launch vehicle: Titan 2
Launch: 16 March 1966 (Gemini 8)
17 March 1966 (ATDA)
Recovery: 16 March 1966 (Gemini 8)
15 September 1967 (ATDA)
Crew: Neil Armstrong, David Scott

The mission was curtailed after seven orbits (10 hours 42 minutes) because of a short circuit in the Gemini 8 Orbital Attitude Maneuvering System (OAMS), but not before the Gemini 8 spacecraft successfully conducted the first docking in space with the ATDA.

Gemini 9
Launch vehicle: Atlas Agena

Above: The Gemini 9 crew prepares to egress after the Agena target vehicle failed to achieve orbit. **Left:** The interiors of all the Gemini spacecraft were identical. **Below:** The Gemini 8 spacecraft docks with the ATDA.

Launch and re-entry: 17 May 1966 (ATDA only)
Crew: Eugene Cernan, Thomas Stafford (Gemini 9)

A second attempt at a docking in space was aborted when the docking target failed to orbit due to launch vehicle malfunction. The Gemini 9 spacecraft was not launched, and was redesignated Gemini 9A for a rescheduled launch two weeks later.

Above: The Gemini 8 astronauts Armstrong and Scott are helped into their spacecraft.

Below: A fish-eye view of Gemini 7 interior showing the complex array of instruments.

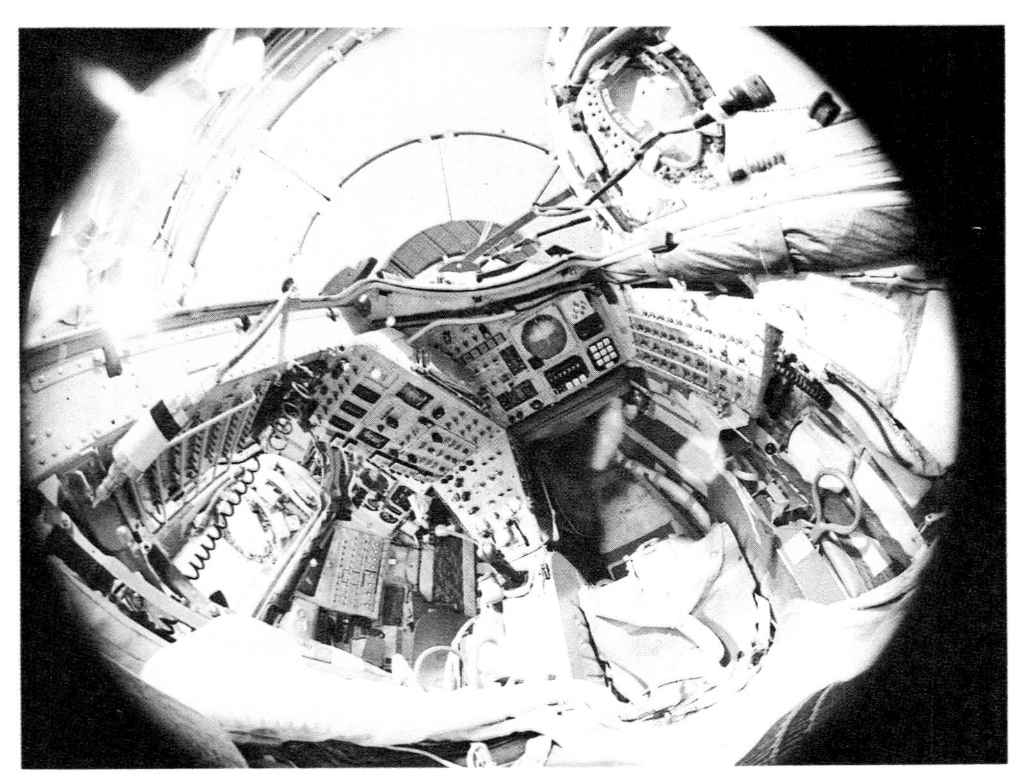

Gemini 9A
Launch vehicle: Titan 2 (Gemini 9A)
Launch: 1 June 1966 (ATDA)
3 June 1966 (Gemini 9A)
Recovery: 6 June 1966 (Gemini 9A)
11 June 1966 (ATDA)

Both the spacecraft and docking target were successfully launched, with Gemini 9A easily completing a rendezvous with the ATDA. Actual docking was not possible, however, because the shroud which covered the docking apparatus during launch was not successfully jettisoned. The spacecraft completed 44 orbits in 72 hours 21 minutes.

Gemini 10
Launch vehicle: Titan 2
Launch: 18 July 1966 (Gemini 10)
18 July 1966 (ATDA)
Recovery: 21 July 1966 (Gemini 10)
Re-entry: 29 December 1966 (ATDA)
Crew: Michael Collins, John Young

Rendezvous and docking were successfully achieved on the fourth of 43 orbits, but 60 percent of the spacecraft's fuel was used up in the maneuver (as opposed to a planned 28 percent) because of a launch error. The crew was recovered after 70 hours 47 minutes.

Gemini 11
Launch vehicle: Titan 2
Launch: 12 September 1966 (Gemini 11)
12 September 1966 (ATDA)
Recovery: 12 September 1966 (Gemini 11)
12 September 1966 (ATDA)
Crew: Charles Conrad, Richard Gordon

The spacecraft completed 44 orbits in 7 hours 17 minutes, and the rendezvous and docking with the ADTA was achieved in 94 minutes on the first orbit.

Gemini 12
Launch vehicle: Titan 2
Launch: 11 November 1966 (Gemini 12)
11 November 1966 (ADTA)
Recovery: 15 November 1966 (Gemini 12)
Re-entry: 23 December 1966 (ADTA)
Crew: Edwin Aldrin, James Lovell

The final spacecraft of the Gemini series completed 59 orbits in 94 hours 34 minutes with Aldrin conducting 5 and one half hours of EVA. Docking was achieved on the third orbit, largely by visual means rather than by radar as with earlier dockings. A solar eclipse was successfully photographed by the Gemini 12 crew on 12 November.

Geodetic Explorer

(See Explorers 29, 36; GEOS)

GEOS

The GEOS (Geodynamic Experimental Ocean Satellites) were one element of NASA's satellite geodesy program, designed to provide precise measurements of the earth's surface and shape, with a mathematical description of its surface and gravity field. The program began with GEOS-A (1965) and GEOS-B (1968) under the umbrella of the Explorer program, and the first independent GEOS launch was in 1976. Related spacecraft included LAGEOS (Laser GEOS) and PAGEOS (Passive GEOS).
(See also LAGEOS, PAGEOS)

GEOS-A
(See Explorer 29)

GEOS-B
(See Explorer 36)

GEOS-C is wired for testing. It carried several geophysical measurements systems.

GEOS-C
Launch vehicle: Delta
Launch: 9 April 1975
Total weight: 750 pounds
Diameter: 53 inches
Height: 32 inches

The GEOS-C was an octagon with a truncated pyramid below the base. Extending toward earth from the pyramid at the base of the octagon was a 235-inch scissor-type boom that held the end mass for the spacecraft's gravity gradient system, providing three-axis stabilization.

This spacecraft was designed by Johns Hopkins Applied Physics Laboratory to demonstrate the feasibility of improved satellite measurements of ocean tides, sea state, gravity and solid-earth dynamics and evaluate new instrumentation. In addition, GEOS-C conducted a satellite-to-satellite tracking experiment with one of the Applications Technology Satellites, ATS-6.

GEOS/ESA
Launch vehicle: Delta 2914
Launch: 20 April 1977
Total weight: 1260 pounds

This spacecraft, while given a GEOS designation, was actually a British-built satellite owned by the European Space Agency (ESA). A cylindrical structure with appendages for several experiments and similar in mission to the GEOS-C, the GEOS/ESA was launched by NASA. The correct orbit was not achieved and the original scientific mission was listed as 'compromised.' A second GEOS/ESA, intended to investigate waves and particles in the magnetosphere, was launched on 14 July 1978.

S X S PROJECT
S X S PROJECT
Aeronutronic
Ford

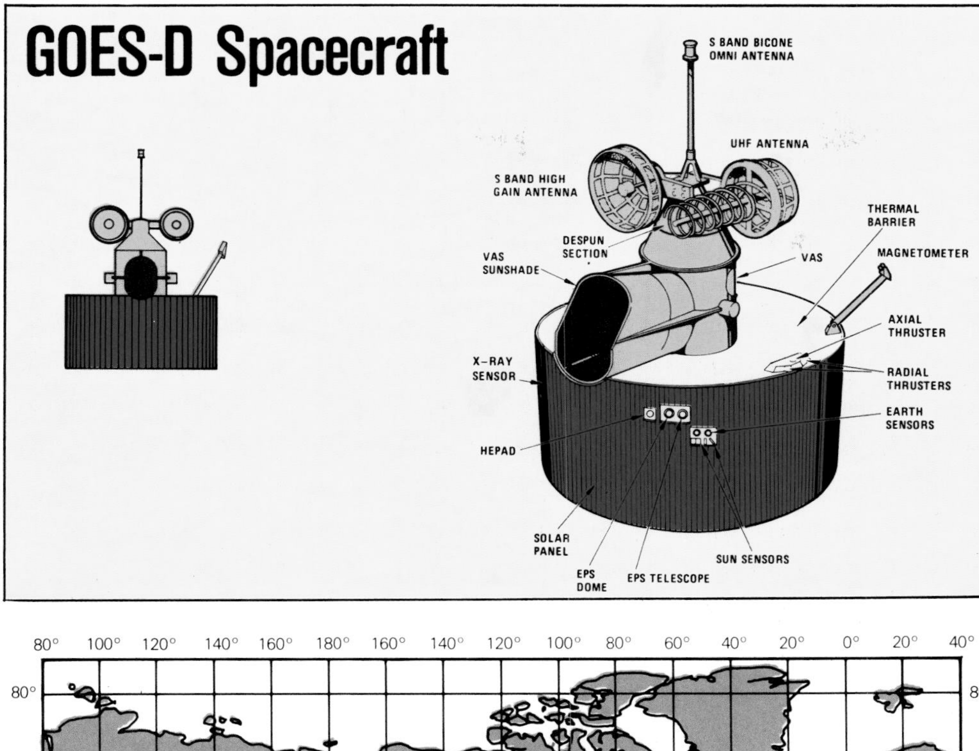

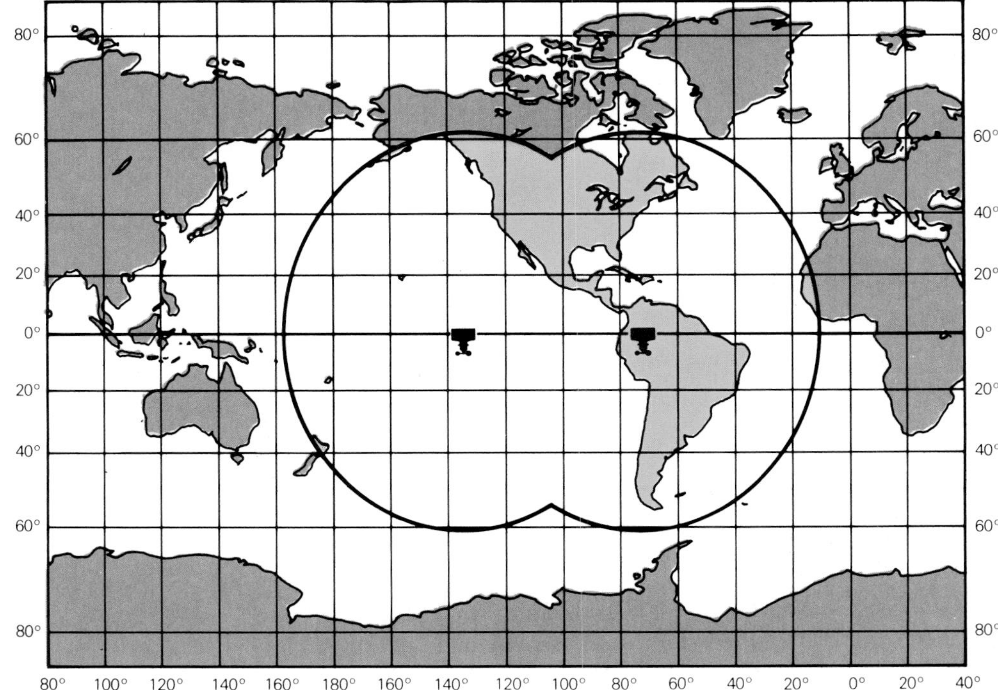

The synchronous-orbit weather satellite GOES-A, which was launched on 16 October 1975.

Global Positioning System

(See NAVSTAR)

GOES

The Geostationary Operational Environmental Satellite (GOES) program was the operational successor to the Synchronous Meteorological Satellite (SMS) program of the National Oceanic and Atmospheric Administration (NOAA). SMD-A (SMS-1) and SMS-B (SMS-2) had been the developmental satellites leading up to the GOES program, and GOES-A also carried the designation SMS-C. The first GOES spacecraft were produced by Ford Aeronutronic, and they were acquired and launched by NASA. After they were placed into orbit, operational control of the GOES satellites was turned over to NOAA, who used them to provide meteorological data on worldwide weather phenomena for improved forecasting and prediction, as well as for observations of ocean currents, monitoring river-water levels and other meteorological evaluations. The spacecraft provided pictures of approximately one fourth of the globe at 30-minute intervals day and night with a visible/infrared spin-scan radiometer. The GOES spacecraft are occasionally confused with the dissimilar GEOS spacecraft because of their similar abbreviations. (See also Synchronous Meteorological Satellites)

GOES-C is encapsulated in its payload fairing.

GOES-A (SMS-C)
Launch vehicle: Delta 2914
Launch: 16 October 1975
Total weight: 1375 pounds at launch
650 pounds in orbit
Diameter: 75 inches
Height: 106 inches

The GOES-A had a cylindrical main body with solar arrays that provided 200 watts of power. Thirty 2-inch magnetometers extended from the end of the cylinder. An apogee kick motor attached to the spacecraft was ejected after synchronous orbit was achieved.

GOES-B (GOES/NOAA)
Launch vehicle: Delta 2914
Launch: 16 June 1977
Specifications and description: Same as GOES-A

GOES-C (GOES-3)
Launch vehicle: Delta 2914
Launch: 16 June 1978
Specifications and description: Same as GOES-A

GOES-D (GOES-4)
Launch vehicle: Delta 3914

Launch: 9 September 1980
Total weight: 1840 pounds at launch
874 pounds in orbit
Diameter: 85 inches
Height: 138 inches
Shape: Cylindrical with solar panels bonded to the entire exterior surface

The GOES-D was the first US spacecraft capable of near-continuous observation of atmospheric water vapor and temperature. The scientific data collected was used in the study of severe storms and storm-spawned phenomena such as hail, flash floods and tornadoes.

GOES-E (GOES-5)
Launch vehicle: Delta 3914
Launch: 22 May 1981
Total weight: 1841 pounds at launch
875 pounds in orbit after firing and ejecting its self-contained solid motor
Diameter: 81 inches
Height: 123 inches

Like GOES-D, this spacecraft was equipped with a visible/infrared spin-scan radiometric atmospheric sounder (VAS), which observed visible light and infrared images of cloud formations and motion, and, on command, temperature variations with light in the atmosphere. It could also map the distribution of water vapor in the atmosphere.

GOES-F (GOES-6)
Launch vehicle: Delta 3914
Launch: 28 April 1983
Specifications and description: Same as GOES-E

Gravity Probe A

Launch vehicle: Scout
Launch and re-entry: 18 June 1976
Total weight: 224 pounds

The role of this spacecraft, launched from Wallops Island on a suborbital mission, was to conduct a scientific test of Einstein's theory of relativity.

Hawkeye
(See Explorer 52)

HCMM (Heat Capacity Mapping Mission)
(See Applications Explorer Mission A)

HEAO

The HEAO (High Energy Astronomy Observatory) program spacecraft were designed to survey and map X-ray sources throughout the celestial sphere; measure low-energy gamma ray flux; investigate astronomical phenomena such as black holes, quasars and pulsars; and provide an all-sky survey of 'soft' X-rays (0.1 KeV) through high-energy (10 MeV) gamma rays with accurate location fixes on their sources. The primary spacecraft contractor was the TRW Defense and Space Systems Group.

Far left: **GOES-D to -F were built by Hughes. *Below left:* HEAO-C studied cosmic rays. *Below:* HEAO-B enclosed in its payload shroud.**

HEAO-C is prepared for launch into earth orbit.

HEAO-A (HEAO-1)
Launch vehicle: Atlas Centaur
Launch: 12 September 1977
Re-entry: 15 March 1979
Total weight: 7000 pounds (including 2600 pounds of scientific payload)

A 10-foot 6-inch-high hexagonal experiment module was built on top of HEAO's octagonal spacecraft equipment module. A rectangular solar array deployed from the top side of the experiment module brought the total height to 20 feet.

HEAO-B (HEAO-2)
Launch vehicle: Atlas Centaur
Launch: 13 November 1978
Re-entry: 25 March 1982
Total weight: 6936 pounds
Diameter: 100 inches (experiment module)
91 inches (SEM)
Height: 226 inches (experiment module)
33 inches (SEM)

The overall height of the spacecraft was 22 feet. The experiment module was eight sided and the SEM was an octagon-shaped prism.

This spacecraft carried the largest X-ray telescope ever built, with a mirror nearly 23 inches wide and a focal length of nearly 128 inches. It was designed to study some of the more puzzling objects in the universe such as pulsars, neutron stars, black holes, quasars, radio galaxies and supernovas.

HEAO-C (HEAO-3)
Launch vehicle: Atlas Centaur
Launch: 20 September 1979
Re-entry: 7 December 1981
Total weight: 6389 pounds

A boxlike experiment module was built on top of a hexagonal spacecraft equipment module. Two large solar arrays and one smaller solar array were attached to one side of the experiment module, which provided HEAO-3 with 415 watts of electrical power. A conical omnidirectional antenna was atop the spacecraft.

The last HEAO mission carried a high-resolution gamma ray spectrometer consisting of four high-purity germanium detectors, surrounded by a cesium-iodide detector that served as a shield and collimater; a cosmic ray experiment consisting of five Cerenkov detectors and a heavy nuclei experiment to determine the energy and charge of individual particles.

Helios

The Helios program was a joint US-German project managed by NASA and the German Ministry for Research and Technology. The spacecraft were produced in Germany by Messerschmitt Boelkow Blohm. The objective of the program was to study the sun from an orbit nearer the center of the solar system than any previously attempted. Both of the spacecraft were launched from the Kennedy Space Center at Cape Canaveral, Florida. The closest approach to the sun was 28 million miles for Helios A and 27 million miles for Helios B.

Helios A
Launch vehicle: Titan Centaur (Titan 3E)
Launch: 10 December 1974
Total weight: 815 pounds

The main body of Helios A was a central 5.7-foot-diameter, 16-sided experiment compartment that was 1.8 feet high. Conical solar arrays extended from each end of the compartment, giving the spacecraft its spool-shaped appearance. Spool height was 7 feet; an antenna mast added another 6.7 feet to the vehicle's height. Two magnetometer booms with a tip-to-tip length of 105 feet were deployed from the sides of the main body. The central compartment housed 11 experiments with a combined weight of 158 pounds.

Helios B
Launch vehicle: Titan Centaur (Titan 3E)
Launch: 15 January 1976
Total weight: 826 pounds in orbit
Diameter: 5.7 feet
Height: 1.8 feet

Helios B consisted of a 16-sided experiment compartment, with conical solar arrays extended from each end of the compartment that gave the spacecraft its spool-shaped appearance. Spool height was 7 feet, and an antenna mast added another 6.7 feet to the vehicle's height. The two conical arrays provided 240 watts of power at its aphelion. Two magnetometer booms with a tip-to-tip length of 105 feet were deployed from the sides of the main body. The central compartment housed 11 experiments with a combined weight of 158 pounds.

HS 376

Launch vehicle: Ariane, Delta or space shuttle
Launch: See Galaxy, SBS, Telstar and Westar
Launch weight: 7300 pounds (average)
Geostationary orbital weight: 1300 pounds (average)
Diameter: 85 inches
Height: 108 inches stowed
260 inches deployed
Solar cells: 14,000 K7 (19.7 mw per C2)
Transponders: 10 to 24 (depending on customer)
Frequencies available: 4-6 GHz (C band) 12-14 GHz (Ku band)
Power plant: Thiokol Star 30 and Star 48 perigee kick motor (PKM)

The Hughes Aircraft HS 376 program includes both a spacecraft and a marketing idea. The idea is for a versatile and powerful communications satellite that is adaptable to a wide variety of customer requirements. Versions of the basic HS 376 can deliver voice, video and/or facsimile transmissions in C band or Ku band or a combination of the two. They are designed to be capable of a 10-year lifespan by virtue of increasing onboard fuel and other technical innovations. Their versatility extends to a choice of launch vehicle. The HS 376 can be launched by NASA's Delta, the European Ariane class vehicle or by the space shuttle.

On 11 November 1982, an HS 376 became the first spacecraft launched by the space shuttle. This HS 376, owned by Satellite Business Systems, carried the designation SBS-3. It was followed three days later by another HS 376, Telesat of Canada's Anik C-3. Hughes Aircraft has taken orders for over 30 of the spacecraft, with customers including AT&T (US), Aussat (Australia), Embratel (Brazil), Hughes Communications (US), the Mexican Secretariat of Communications and Transportation, Perumtel (Indonesia), SBS (US), Telesat (Canada) and Western Union (US). The satellites are particularly important to Indonesia, whose HS 376s (Palapa) serve a vast archi-

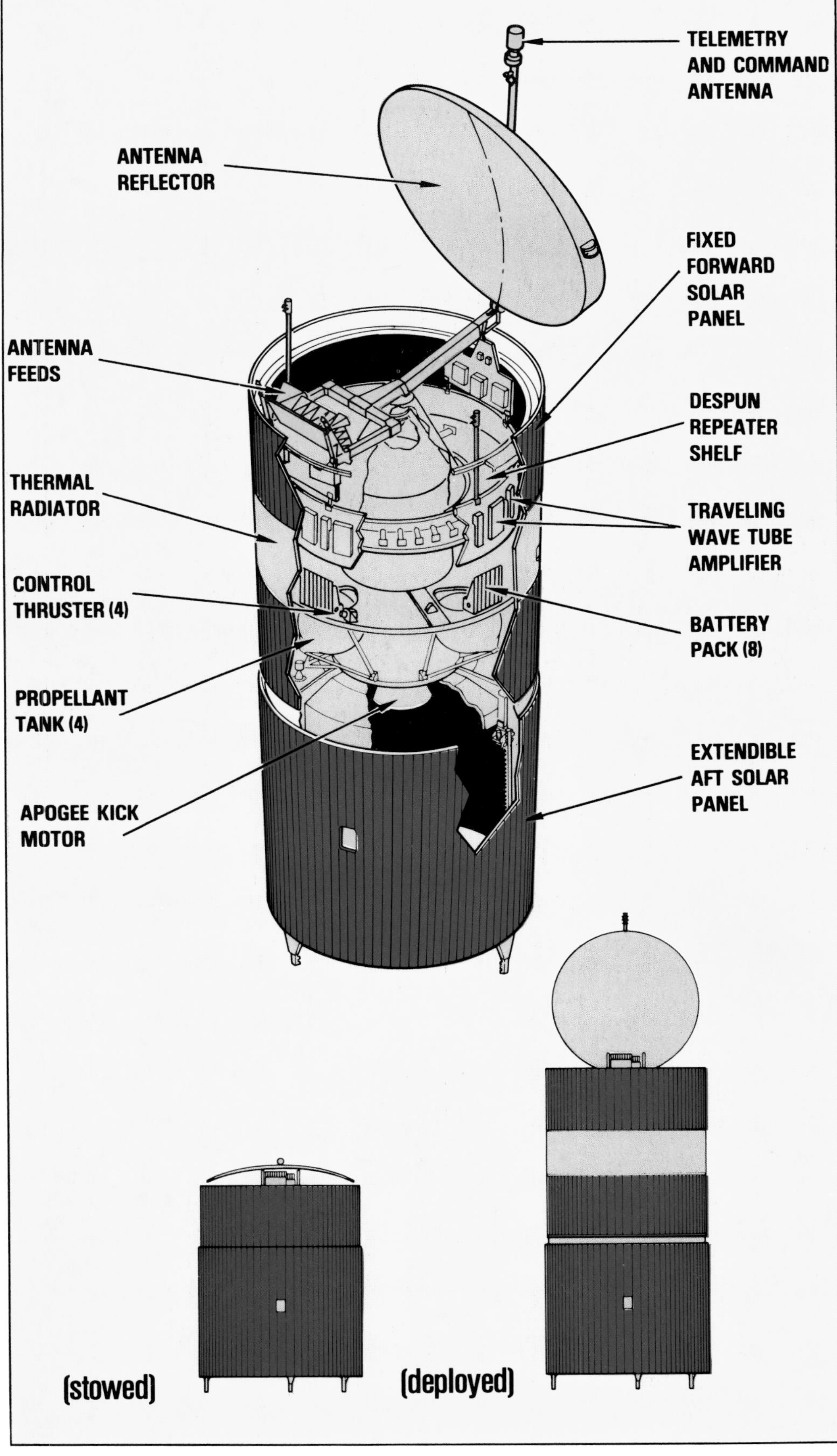

Hughes HS 376: Deployment from the Space Shuttle

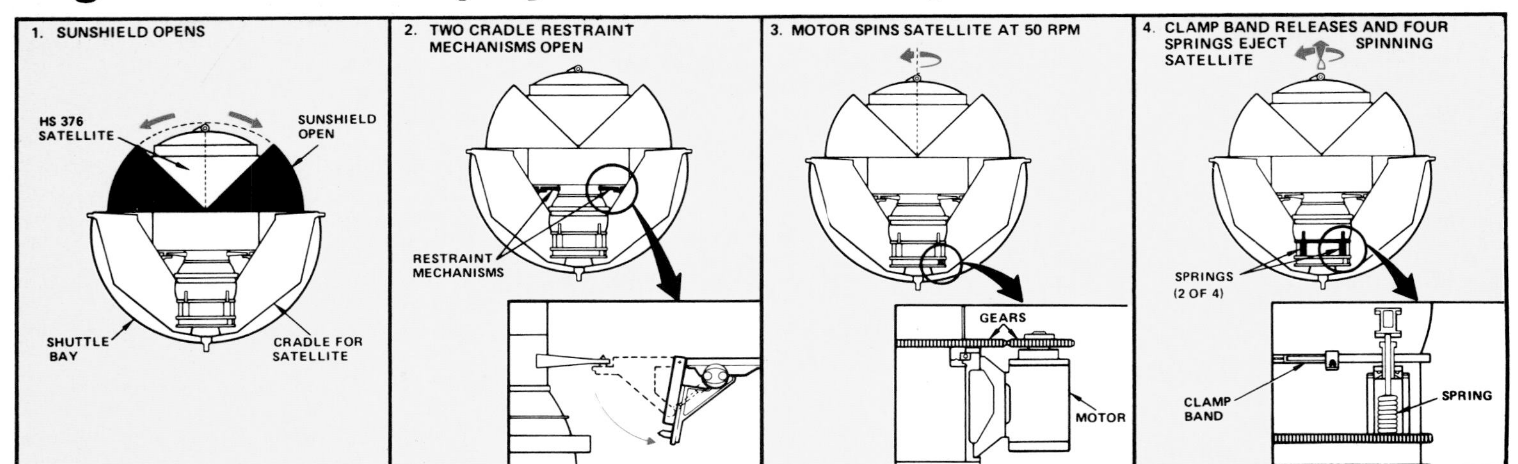

Hughes HS 376: Orbital positioning

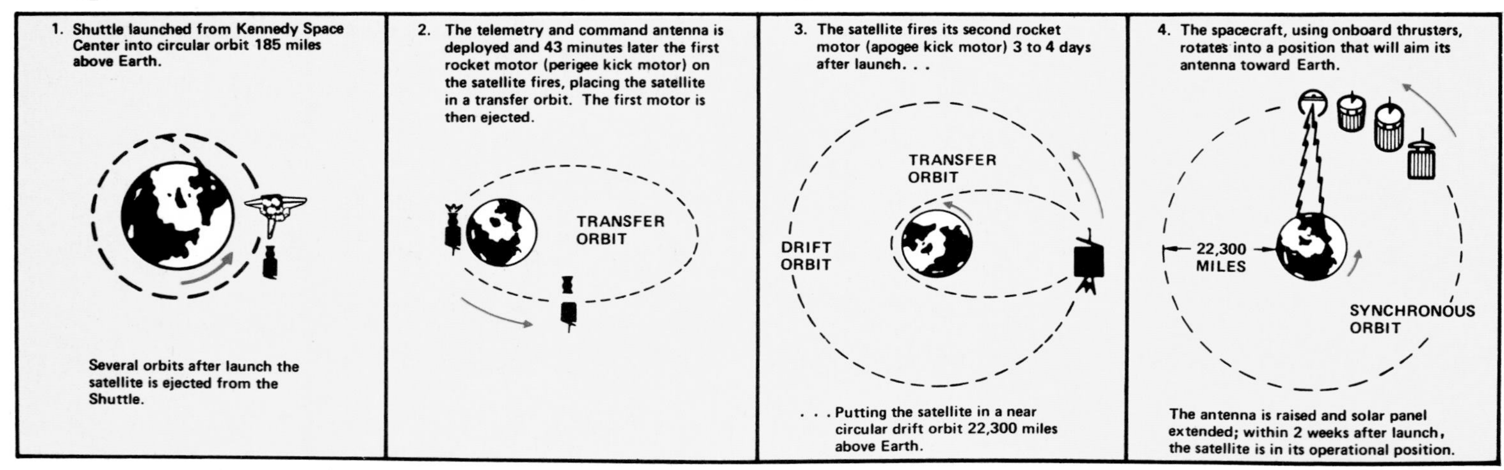

pelago where other types of communications would be prohibitively expensive.

The HS 376 is a compact 108 inches tall when stowed, using only 14 percent of the shuttle's payload bay. However, it folds out to 260 inches, including a 72-inch antenna, when it is fully deployed.

Transmit and receive beams were created by the shared aperture grid antenna with two polarization-selective surfaces. The front surface was sensitive to horizontally polarized beams and the rear surface to vertically polarized beams. Separate microwave feed networks were used for the different polarizations.

(See also Galaxy, SBS, Telstar, Westar)

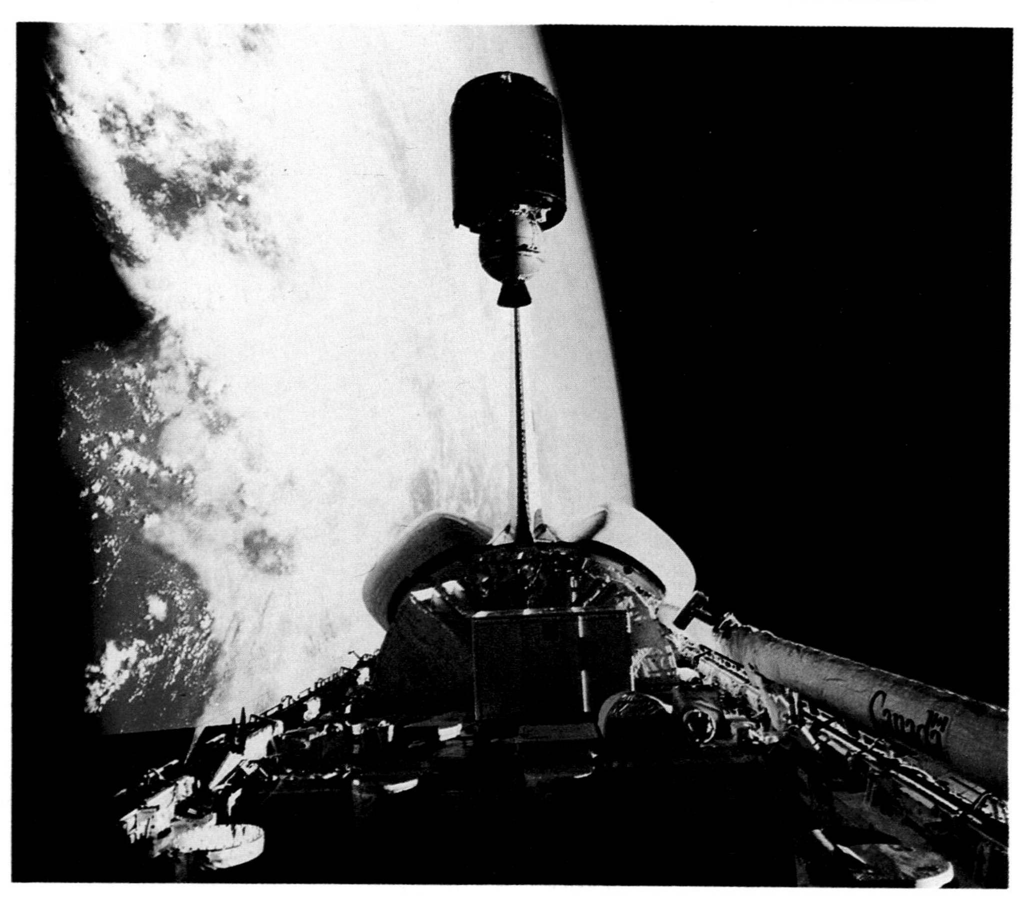

IDSCS

(See DSCS 1)

IMEWS

Launch vehicle: Titan 3C
Launch: 6 November 1970 (IMEWS 1)
5 May 1971 (IMEWS 2)
1 March 1972 (IMEWS 3)
Total weight: 1804 pounds

***Above and right:* Engineers prepare Intelsat 1 for launch. A Hughes HS 376 (*below left*) about to clear the vertical stabilizer of *Challenger*.**

The spacecraft consisted of two flat drumlike cylinders connected to an elongated cylinder by a truncated cone. Sensors and antennae were deployed from the sides of the spacecraft as were a cross-shaped solar array spanning 273 inches.

The IMEWS (Integrated Missile Early Warning Satellite) program, a successor to the Midas program, was a military program to detect launches of hostile ICBMs or SLBMs from geosynchronous orbit in outer space by means of infrared sensors. The IMEWS spacecraft, with a five-year operational life, have successfully provided immediate detection of Soviet and Chinese ICBM tests. The IMEWS 1 spacecraft did not go on station because of launch vehicle failure. IMEWS 2 was placed over the Indian Ocean and IMEWS 3 over Central America.

Intelsat

Television broadcast frequency: 6390 Mc (from US)
6301 Mc (from Europe)
4801 Mc (to US)
4161 Mc (to Europe)
Primary contractor: Hughes

The Intelsat communications satellites were developed by various contractors for the Communications Satellite Corporation (COMSAT) and launched by NASA. Owned and operated by COMSAT in cooperation with the International Telecommunications Satellite Consortium (Intelsat), these were designed to transmit voice, data, facsimile and telegraphy as well as television. The series of over two dozen spacecraft is composed of five blocks, or types, of spacecraft ranging from the tiny Intelsat Block 1 weighing 85 pounds to the huge Intelsat Block 5 spacecraft that weigh over two tons. The blocks are subdivided into individual spacecraft flights, which are designated with 'F' prefixes. The first Intelsat spacecraft, nicknamed *Early Bird*, was launched in April 1965 and began commercial service on 28 June 1965. *Early Bird* was the only Intelsat Block 1 spacecraft.

Intelsat 1
Launch vehicle: Thrust Augmented Delta (TAD)
Launch: 6 April 1965 (Intelsat 1 F-1, *Early Bird*)
Total weight: 85 pounds
Height: 23 inches
Diameter: 28 inches
Shape: Cylindrical
Shell composition: Aluminum and magnesium covered with solar cells

Intelsat 2
Launch vehicle: Delta (DSV-3E)
Launch: 26 October 1966 (F-1, *Lani Bird*)
11 January 1967 (F-2)
23 March 1967 (F-3)
28 September 1967 (F-4)
Total weight: 357 pounds at launch
192 pounds in orbit
Diameter: 56 inches
Height: 26.5 inches
Shape: Cylindrical
Shell composition: Sides faced with solar cells, the ends with thermal shields
Frequency: 125 Mc with 240 two-way voice channels
Primary contractor: Hughes

Intelsat 3
Launch vehicle: Delta
Launch: 19 September 1968 (F-1)

Intelsat 2 (*above*) suspended above Los Angeles and 3 (*above right*) undergo final checks.

The Intelsat 4As (*right*) and the Intelsat 4s (*below*) make up a global satellite system.

(destroyed 68 seconds after launch due to control-system failure)
18 December 1968 (F-2)
5 February 1969 (F-3)
22 May 1969 (F-4)
26 July 1969 (F-5) (failure at launch)
14 January 1970 (F-6)
22 April 1970 (F-7)
23 July 1970 (F-8) (failure at launch)
Total weight: 642 pounds (through F-2)
322 pounds (through F-5)
342 pounds (through F-8)
(weights with propellants)
Diameter: 56 inches
Height: 41 inches (78 inches with antenna deployed)
Shell composition: Magnesium
Shape: Cylindrical
Receiving frequencies: 5930-6429 MHz (through F-2)
5920-6420 MHz (through F-4)
5930-6420 MHz (through F-8)
Transmission frequencies: 3705-4195 MHz (through F-2)
3695-4195 MHz (through F-8)
Number of channels: 1200
Primary contractor: TRW

The exterior of the craft was covered with solar cells. The aft and forward surfaces were enclosed except for slight extensions of the apogee motor skirt aft and mechanically despun communications

antenna, shaped like a cone with a base cut at 45 degrees that extended 33.6 inches forward. Mounted around the base of the communications antenna was a doughnut-shaped omnidirectional telemetry and command antenna. Two axial thrusters were mounted aft, and two radial thrusters were mounted at the side. Earth and sun sensors were also mounted at the side. Supportive subsystem equipment was mounted inside on the circular aluminum honeycomb platform. Spin was stabilized at 90 rpm.

Intelsat 4
Launch vehicle: Atlas Centaur
Launch: 22 May 1975 (F-1)
25 January 1971 (F-2)
19 December 1971 (F-3)
22 January 1972 (F-4)
13 June 1972 (F-5)
20 February 1975 (F-6) (launch vehicle failure)
23 August 1973 (F-7)
21 November 1974 (F-8)
(Note: F-1 and F-6 were not launched in sequence)
Total weight: 1600 pounds (F-1, F-6)
1559 pounds (F-2, F-3)
1587 pounds (F-5)
1500 pounds (F-7, F-8)
Diameter: 93.7 inches (F-1, F-6)
94 inches (F-2, F-3)
96 inches (F-4 through F-8, except F-6)
Height: 111 inches (204 to 210 inches with antennae deployed)
Shape: Cylindrical
Television frequency bandwidth: 432 MHz (36 MHz for F-1, F-6)
Receiving frequency: 5932-6418 MHz
Transmission frequency: 3707-4192 MHz
Voice and data channels: 4000 (F-1, F-6)
9000 (F-2, F-3)
6000 (F-4 through F-8, except F-6)
Televison channels: 12
Primary contractor: Hughes

Intelsat 4A
Launch vehicle: Atlas Centaur
Launch: 25 September 1975 (F-1)
29 January 1976 (F-2)
7 January 1978 (F-3)
26 May 1977 (F-4)
29 September 1977 (F-5)
31 March 1978 (F-6)
(Note: Intelsat 4A F-3 was launched out of sequence)
Total weight: 3333 pounds at launch
1750 pounds in orbit
Diameter: 94 inches;
53 inches (each of two transmission antenna dishes)

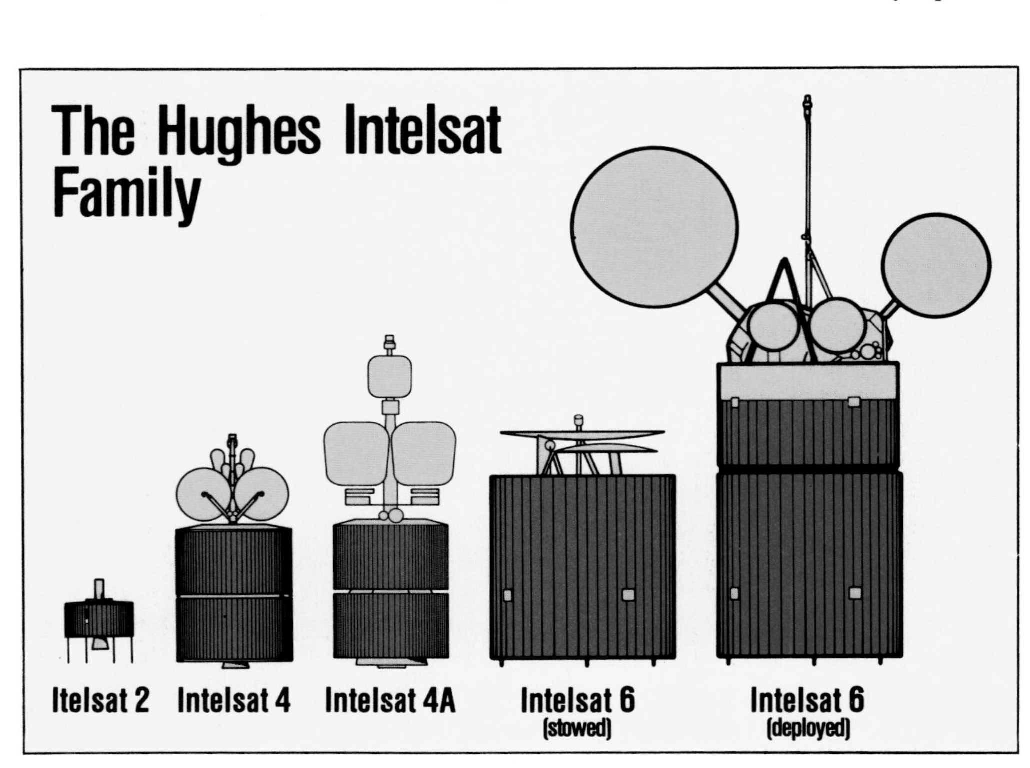

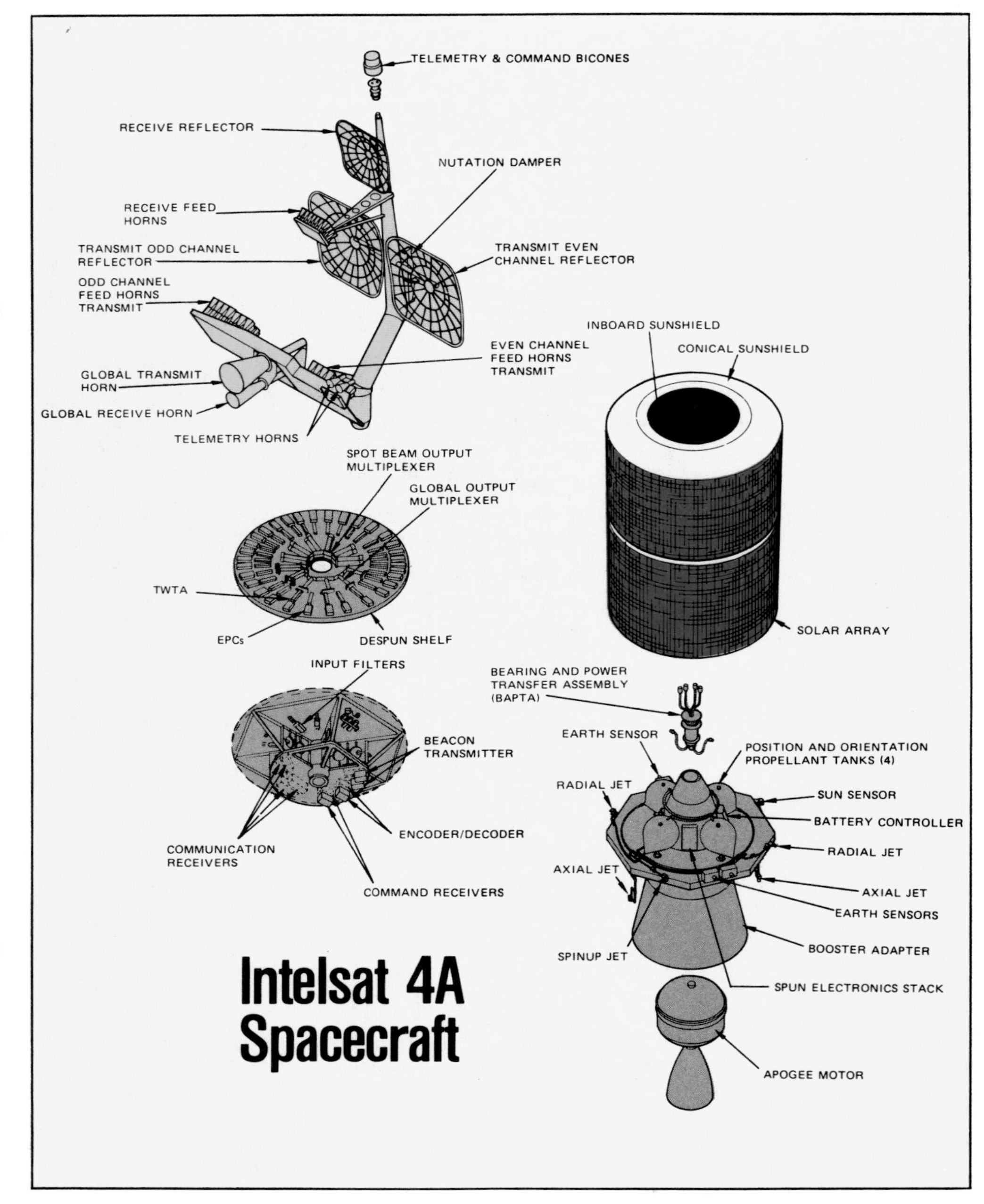

Far right: **Intelsat 4A had a spot-beam antenna.**

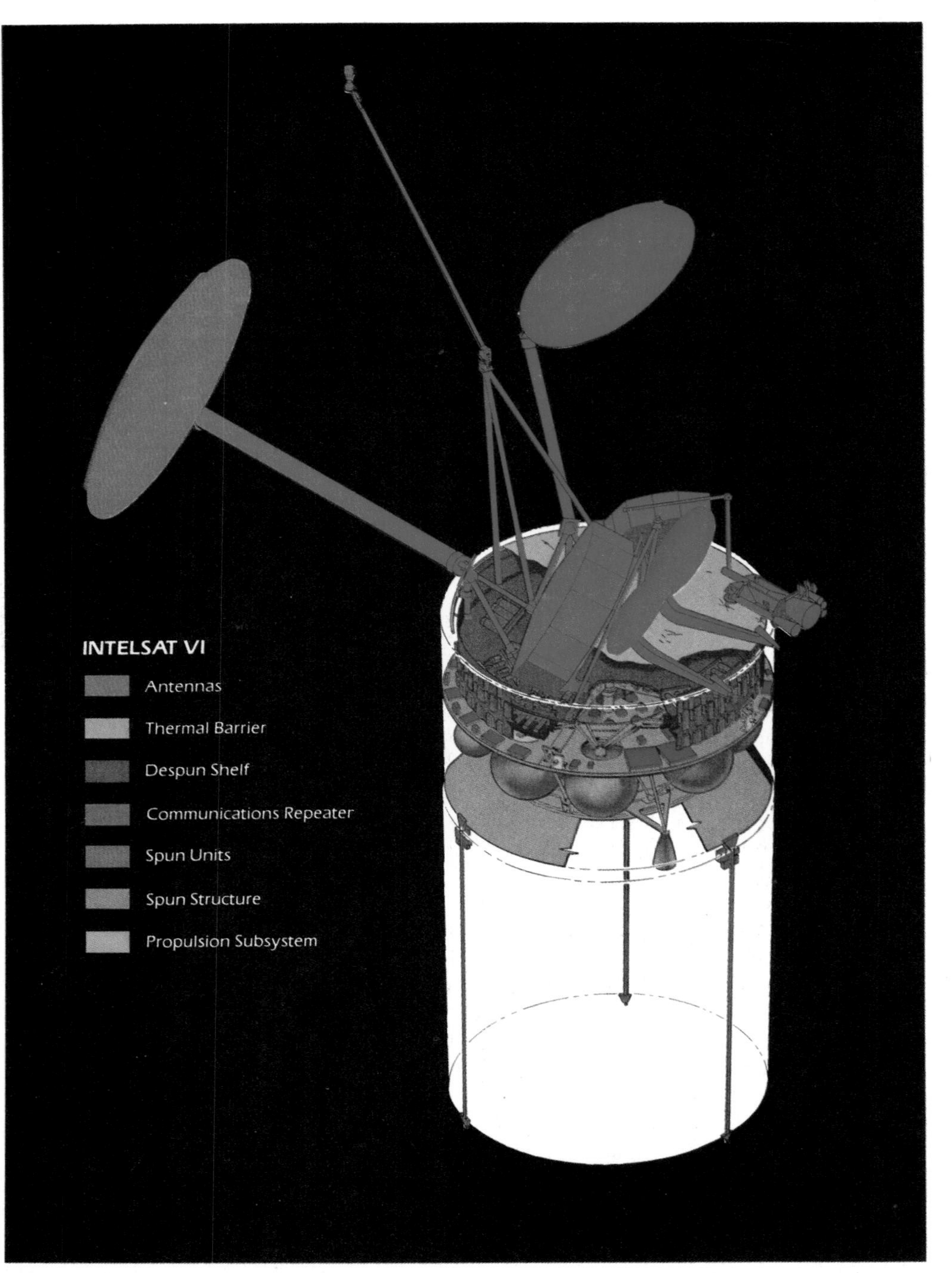

Height: 111 inches (267 inches with antennae deployed)
Shape: Cylindrical
Frequency bandwidth: 36 MHz (each of 20 transponders)
Voice and data channels: 6250
Television channels: 2
Primary contractor: Hughes

Intelsat 5
Launch vehicle: Atlas Centaur (AC-54)
Launch: 23 May 1981 (F-1)
6 December 1980 (F-2) (B)
15 December 1981 (F-3) (C)
4 March 1982 (F-4)
28 September 1982 (F-5)
19 May 1983 (F-6)
December 1983 (5A F-1)
9 June 1984 (5G)
Total weight: 4251 pounds at launch

Intelsat 6 is being built by an international team. *Below:* The solar panel (2), transmit reflector (3) and antennae (4) are deployed.

Width: 69.6 inches (267 inches with solar panels deployed)
Height: 79.2 inches (252 inches with antennae deployed)
Depth: 64.8 inches
Frequency band: 16/11 GHz and 6/4 GHz
Voice and data channels: 12,000
Primary contractor: Ford Aerospace and Communications Corp

The three-axis body of the Intelsat 5 spacecraft was a modular construction stabilized with sun and earth sensors and the momentum wheel. Its winglike sun-oriented solar array panels produced a total of 1241 watts of electrical power after seven years in orbit.

Intelsat 6
Launch vehicle: Space shuttle/Ariane 4
First launch: 1986
Total weight: 3918 pounds
Diameter: 144 inches
Height: 210 inches (stowed)
468 inches (deployed)
Shape: Cylindrical
Frequency spectrum: 6/4 GHz (C band)
14/11 GHz (K band)
Voice and data channels: 33,000
Television channels: four

International Ultraviolet Explorer

Launch vehicle: Delta
Launch: 26 January 1978
Total weight: 1476 pounds (with apogee motor)
Diameter: 50.7 inches (spacecraft body)
Height: 167.6 inches (spacecraft body)
Width: 167.7 inches (with solar panels deployed)
Shape: Octagonal

The International Ultraviolet Explorer (IUE) project was a cooperative joint effort between NASA, ESA and Britain's Science Research Council, with NASA's Goddard Space Flight Center primarily responsible for the design and production of the spacecraft. It was designed to examine the ultraviolet spectral region between 1150 and 3200 Angstroms, a region invisible from earth, and emissions of hydrogen, helium, carbon and oxygen. Studies were also conducted of the solar system and distant objects such as quasars, pulsars and black holes.

Interplanetary Monitoring Platform (IMP)

(See Explorers 18, 21, 28, 33, 34, 35, 41, 43, 47, 50)

Ionosphere Explorer

(See Explorer 20)

IRAS

Launch vehicle: Delta
Launch: 25 January 1983
Total weight: 2249 pounds

The IRAS (Infrared Astronomical Satellite) project is a joint effort between NASA and the Netherlands, with British participation. The spacecraft itself was developed by NASA's Jet Propulsion Laboratory in conjuction with Ball Aerospace in the US and Fokker in the Netherlands. Launched from the Western Space and Missile Center at Vandenberg AFB, the spacecraft was designed to conduct scientific infrared studies of deep space.

ISEE

The ISEE (International Sun-Earth Explorer) program spacecraft were designed to study the ambiguities associated with the motion of the boundaries of space near the earth. The three spacecraft provided data similar to one another and data similar to planetary probes. The results gave 117 scientists representing 35 universities in 10 nations a better idea of how the sun controls the earth's immediate space environment.

Launch vehicle: Delta
Launch: 22 October 1977 (ISEE 1, ISEE 2)
12 August 1977 (ISEE 3)
Total weight: 1034 pounds
Diameter: 68.4 inches
Height: 63.6 inches
Shape: 16-sided polyhedron
Primary contractors: NASA Goddard (ISEE 1, 3)
Dornier/European Space Technology Center (ISEE 2)

The main body of the ISEE consisted of a 33-inch conical center tube. Solar cells, mounted on the sides of the spacecraft, provided power for the experiments and transmission.

ISEE 3 streaks 72 miles above the lunar surface, catapulted by the moon's gravity to an encounter with Comet Giacobini-Zinner.

ISIS

Launch vehicle: Delta
Launch: 30 January 1969 (ISIS A)
31 March 1971 (ISIS B)
Total weight: 532 pounds (ISIS A)
582 pounds (ISIS B)
Diameter: 50 inches
Height: 42 inches (ISIS A)
48 inches (ISIS B)

The ISIS (International Satellites for Ionospheric Studies) program was a cooperative venture between NASA and the Canadian Defense Research Board. The spacecraft were built by NASA's Goddard Space Flight Center and RCA Victor Ltd of Montreal, Canada. The two operational ISIS spacecraft were successors to the Canadian Alouette 2 spacecraft that had been launched in 1965 (simultaneously with the US Explorer 31) and that had also borne the secondary designation ISIS-X. The objective of the ISIS A was to study the depth, composition and behavior of the upper ionosphere for ionospheric disturbances and polar blackouts, with the goal of improving long-range radio transmission. The objective of ISIS B was to measure daily and seasonal fluctuations in the electron density of the upper ionosphere, and to study radio and cosmic noise emissions, as well as the interaction of energy particles with the ionosphere.

Like Alouettes 1 and 2, ISIS was spherical in configuration and its surface covered by 11,136 solar cells. The chief difference betwen the Alouettes and ISIS was the added capability on ISIS for onboard data storage. Two extendable antennae, 240 and 62 feet long, sounded the upper ionosphere; four telemetry antennae projected from the base; beacon antennae were mounted around the satellite equator; and two antenna-like booms supported probes for several onboard experiments. Attitude control was also newly incorporated into this design, maintained by a spin-stabilization system including a three-axis magnetometer and solar-aspect sensor. Magnetic torquing controls spun at a rate of within one to three rpms.

ITOS

The ITOS, or Improved TIROS Operational System, was a series of weather satellites that evolved out of the TIROS and Environmental Science Services Administration (ESSA) programs. These spacecraft were produced by RCA, the primary contractor, and operated by ESSA or NOAA. Most of the ITOS spacecraft also carried designations. (See NOAA, TIROS)

ITOS-1 (TIROS-M)
Launch vehicle: Delta N
Launch: 23 January 1970
Total weight: 682 pounds
Diameter: 40 inches (spacecraft body)
Height: 48 inches (spacecraft body)
Width: 14 feet (with solar panels deployed)
Shape: Rectangular
Surface composition: Multilayer foil insulation
Systems: 2 advanced vidicon camera systems (AVCS)
2 automatic picture transmission systems (APT)
1 flat plate radiometer (FPR)
1 solar proton monitor (SPM)

ITOS-1 was successfully placed into orbit and turned over to ESSA on 15 June 1970.

ITOS-A (NOAA 1)
Launch vehicle: Delta N
Launch: 11 December 1970
Specifications and description: Same as ITOS 1
Systems: Same as ITOS 1, plus a Cylindrical Electrostatic Probe Experiment (CEPE) carried piggyback

ITOS-B (NOAA 2)
Launch vehicle: Delta N
Launch: 21 October 1971
Specifications and description: Same as ITOS 1
Systems: Same as ITOS 1

ITOS-B failed to achieve orbit due to launch vehicle failure.

ITOS-C
(Designation not assigned)

ITOS-D (NOAA 2)
Launch vehicle: Delta 300
Launch: 15 October 1972
Total weight: 736 pounds
Specifications and description: Same as ITOS 1 (except body was 1 inch taller)
Systems: 1 flat plate radiometer (FPR)
1 solar proton monitor (SPM)

Left: ITOS-A was designated NOAA-1 when it entered orbit 910 miles above the earth.

Above: ITOS-F was successfully launched from the Western Test Range in November 1973.

ITOS/NOAA

Spacecraft	Launch	Duration (days)	Weight (lb)	Systems
ITOS-1 (TIROS M)	23 Jan 1970	510	682	AVCS, APT, FPR, SPM
ITOS-A (NOAA 1)	11 Dec 1970	252	682	AVCS, APT, FPR, SPM
ITOS-B (NOAA 2)	21 Oct 1971	0	682	AVCS, APT, FPR, SPM
ITOS-C	–	–	–	–
ITOS-D (NOAA 2)	15 Oct 1972	837	736	FPR, SPM, VTPR, VHRR
ITOS-E (NOAA 3)	16 July 1971	0	750	FPR, SPM, VTPR, VHRR
ITOS-F (NOAA 3)	6 Nov 1973	1029	750	FPR, SPM, VTPR, VHRR
ITOS-G (NOAA 4)	15 Nov 1974	1463	750	FPR, SPM, VTPR VHRR
ITOS-H (NOAA 5)	29 July 1976	1036	750	FPR, FPM, VTPR, VHRR
NOAA 6	27 June 1979	*	1590.6	AVHRR, ASS, SEM
NOAA 7	29 May 1980	*	1590.6	AVHRR, TOVS, SEM, SPM, TIP, MIR, CSU
NOAA 7C	23 June 1981	*	1590.6	AVHRR, TOVS, SEM, SPM, TIP, MIR, CSU
NOAA 8	28 Mar 1983	*	1590.6	AVHRR, SEM
NOAA 9	9 Dec 1984	*	1590.6	AVHRR, SEM

*Duration of still-operating spacecraft expected to match earlier durations
(See also ESSA, NOAA, TIROS)

ITOS-G was launched with AMSAT-OSCAR-B.

1 vertical temperature profile radiometer (VTPR)
1 very high resolution radiometer (VHRR)
No cameras

ITOS-D was a successful replacement for ITOS-B.

ITOS-E
Launch vehicle: Delta 300
Launch: 16 July 1973
Total weight: 750 pounds
Specifications and description: Same as ITOS-D
Systems: Same as ITOS-D

ITOS-E failed to achieve orbit due to launch vehicle failure.

ITOS-F (NOAA 3)
Launch vehicle: Delta 300
Launch: 6 November 1973
Total weight: 750 pounds
Specifications and description: Same as ITOS-D
Systems: Same as ITOS-D

ITOS-F was a successful replacement for ITOS-E.

ITOS-G (NOAA 4)
Launch vehicle: Delta 2310
Launch: 15 November 1974
Total weight: 750 pounds
Specifications and description: Same as ITOS-D
Systems: Same as ITOS-D

ITOS-H (NOAA 5)
Launch vehicle: Delta 2310
Launch: 29 July 1976
Total weight: 750 pounds
Specifications and description: Same as ITOS-D
Systems: Same as ITOS-D

(Note: The subsequent NOAA-designated spacecraft, from NOAA 6 on, had no equivalent ITOS designation. They were part of the TIROS-N series. See also NOAA, TIROS.)

Key Hole

Launch vehicle: Titan 3D/Agena
First launch: 19 December 1976
Total weight: 30,000 pounds (approx)
Diameter: 6.6 feet (approx)
Height: 64 feet (approx)

The Key Hole spacecraft (Code 1010) are among the most sophisticated reconnaissance satellites known to exist in the US arsenal. Developed jointly by the US Air Force and the CIA, the Key Hole spacecraft would likely have remained classified had a junior CIA operative named William Kampiles not sold a KH-11 manual to the Soviets in 1978. Though many of the spacecraft's finer points are still secret, some data has emerged, partly because of the celebrated spy trial that followed Kampiles' arrest.

The KH-11 represents the most sophisticated known development of the Key Hole series. Data and pictures from the KH-11 are digitized and returned to earth stations telemetrically rather than physically in cassettes, as was the case with Big Bird and earlier Key Holes such as KH-9 and the low-altitude KH-8. The KH-8 and KH-9, with their higher-resolution photographic systems, were taken out of production in order to make more money available for production of the KH-11.

Though similar in size and mission to the Big Bird satellites, the KH-11 is more advanced and can remain in space for two years, four times longer than a Big Bird. These Key Hole spacecraft offer broad coverage with a variety of sensors from an elliptical orbit 185 by 275 miles in space. The sensors are extremely sensitive, with a ground resolution of about 10 feet. They include 'multispectral' scanners that can observe both the visible and invisible portions of the spectrum, including infrared bands. Among the alleged capabilities of the KH-11's sensors is the ability to photograph events that took place in the past, by observing residual heat that might remain from a missile launch or troop movements.

Above: A Titan 3 launch vehicle of the type used to launch military recon satellites like KH-11. *Right:* LAGEOS was designed to provide a stable point in the sky to reflect pulses of laser light to help predict earthquakes.

LAGEOS

Launch vehicle: Delta 2914
Launch: 4 May 1976
Total weight: 903 pounds

The LAGEOS (Laser Geodynamics Satellite) is a solid, spherical passive satellite with no moving parts or electronic components. The basic aluminum sphere with a solid brass core was 24 inches in diameter. It carried an array of 426 prisms called cube-corner retroreflectors, giving it the dimpled appearance of a golf ball. Retroreflectors were three-dimensional prisms that reflected the laser beams back to their source, regardless of the angle at which they were received. LAGEOS retroreflectors were made of high-optical-quality fused silica (a synthetic quartz).

The objective of the LAGEOS program was to provide a reference point for laser ranging experiments, such as monitoring of the motion of the earth's tectonic plates, studying the time-varying behavior of the earth's polar positions, maintaining the geodetic reference system and providing a more accurate determination of universal time. Developed by NASA's Marshall Space Flight Center, LAGEOS was the first satellite in the solid earth

dynamics segment of NASA's Earth and Ocean Dynamics Applications Program (EODAP).
(See also GEOS, PAGEOS)

Landsat (ERTS)

Launch vehicle: Delta
Launch: 23 July 1972 (ERTS-A, renamed Landsat 1)
22 January 1975 (Landsat 2)
5 March 1978 (Landsat 3)
16 July 1982 (Landsat 4)
1 March 1984 (Landsat 5)
Total weight: 1965 to 2100 pounds
Diameter: 59.3 inches
Height: 118.6 inches
Width: 13 feet (with solar panels deployed)
Systems: Return beam vidicon (RBV) and multispectral scanner (MSS) sensors provided independent images of earth surface areas 115 miles square

The objective of the ERTS (Earth Resources Technology Satellite) program was to utilize unmanned spacecraft to survey the earth's surface photographically via the visible and infrared portions of the spectrum on a repetitive basis and to investigate the practical commercial and scientific application of the results. The Landsat spacecraft were used to map and

Above: Landsat 4 views of New York City and Death Valley (_far right_) produced by the Thematic Mapper. The butterfly-shaped Landsat satellites (_right_) have had remarkable success photographing the earth.

NASA
PACE
TER
ECTRIC

catalog the earth's natural resources and monitor changing environmental conditions. Not only were spectacular photographs of most of the earth's surface returned, but practical applications were found in the areas of agriculture, forestry and mining. The spacecraft were all developed by RCA in conjunction with NASA's Goddard Spaceflight Center, and launched from the Western Space and Missile Center at Vandenberg AFB.

The Landsat spacecraft design was patterned after the successful Nimbus series of meteorological satellites. They consisted of integrated subsystems that provided power, thermal control and orbit maintenance of the sensor payloads for one year in orbit. Attitude of the spacecraft was controlled with a pointing accuracy of less than 0.7 degrees in roll, yaw and pitch axes. Two wide-band video tape recorders stored up to 30 minutes each of picture information for delayed readout when the spacecraft was in view of a data acquisition station.

(See also Seasat)

Left: Landsat 2 provided coverage of the entire globe for a broad range of studies.
Below: Landsat 3 carried cameras that measured natural resources in total darkness.

Above and right: The widebody Leasat 2 was designed for use with the space shuttle.

Leasat

Launch vehicle: Space shuttle
First launches: 1 September 1984
10 November 1984
Total weight: 2900 pounds
Diameter: 166.1 inches
Height: 169.3 inches (242.5 inches with solar panels deployed)

Developed by Hughes, the Leasat program was designed to serve as a successor to the US Navy FLTSATCOM system of communications satellites. The system is composed of four satellites in geosynchronous orbit, with a solar array generating 1260 watts of power for UHF and SHF transmissions.
(See also FLTSATCOM, Syncom)

Long Duration Exposure Facility

Launch vehicle: Space shuttle
Launch: 8 April 1984
Total weight: 21,400 pounds

Below: The RMS arm releases the Long Duration Exposure Facility into space.

Diameter: 14 feet
Height: 30 feet
Shape: 12-sided cylinder

Deployed by the orbiter *Challenger*, the Long Duration Exposure Facility was developed by the Langley Research Center to gather exposure data during a one-year period in space. It was designed to be recoverable by an STS orbiter.

Lunar Orbiter

The primary objective of the lunar orbiter program was to obtain topographic data in the lunar equitorial region (43°E to 56°W) to help select suitable landing sites for the Surveyor unmanned and the Apollo manned lunar landing missions. The spacecraft, developed by Boeing Aerospace, were noted for their maneuverability, an attribute that greatly enhanced the success of the mission. The primary objective was accomplished by Lunar Orbiter 3, and the fourth and fifth flights were devoted to a further photographic survey of lunar surface features for scientific purposes. Lunar Orbiter 4 surveyed 99 percent of the front of the moon and much of the back side.

The lunar orbiter's 150-pound Eastman Kodak camera system included 88 mm medium-resolution and 24-inch high-resolution lenses set a f5.6 with aperture speeds adjustable to 1/23, 1/50 and 1/100 second. The focal plane was moved during the exposures to prevent blurring. There were 194 to 212 dual-exposure frames exposed on a 200-foot roll (260 feet for Lunar Orbiter 5) of 70 mm 50-243 film. The resulting negatives were video scanned and the results transmitted back to earth stations.
(See also Apollo, Surveyor)

Below: The Long Duration Exposure Facility was launched in 1984 and retrieved in February 1985.

Above: **Artist's concept of Lunar Orbiter 1 shows the camera system and electronic equipment.**

Lunar Orbiter 1
Launch vehicle: Atlas Agena
Launch: 10 August 1966
Readout complete: 13 September 1966
Impact on the moon: 19 October 1966
Total weight: 853 pounds
Span: 222 inches (across antennae booms)
Length: 146 inches (across solar panels)
Height: 66 inches
Shell composition: Main body covered by an aluminized mylar reflective thermal blanket

A low-gain antenna and a parabolic high-gain antenna extended from the base of the truncated cone structure at 180 degrees from each other and four solar

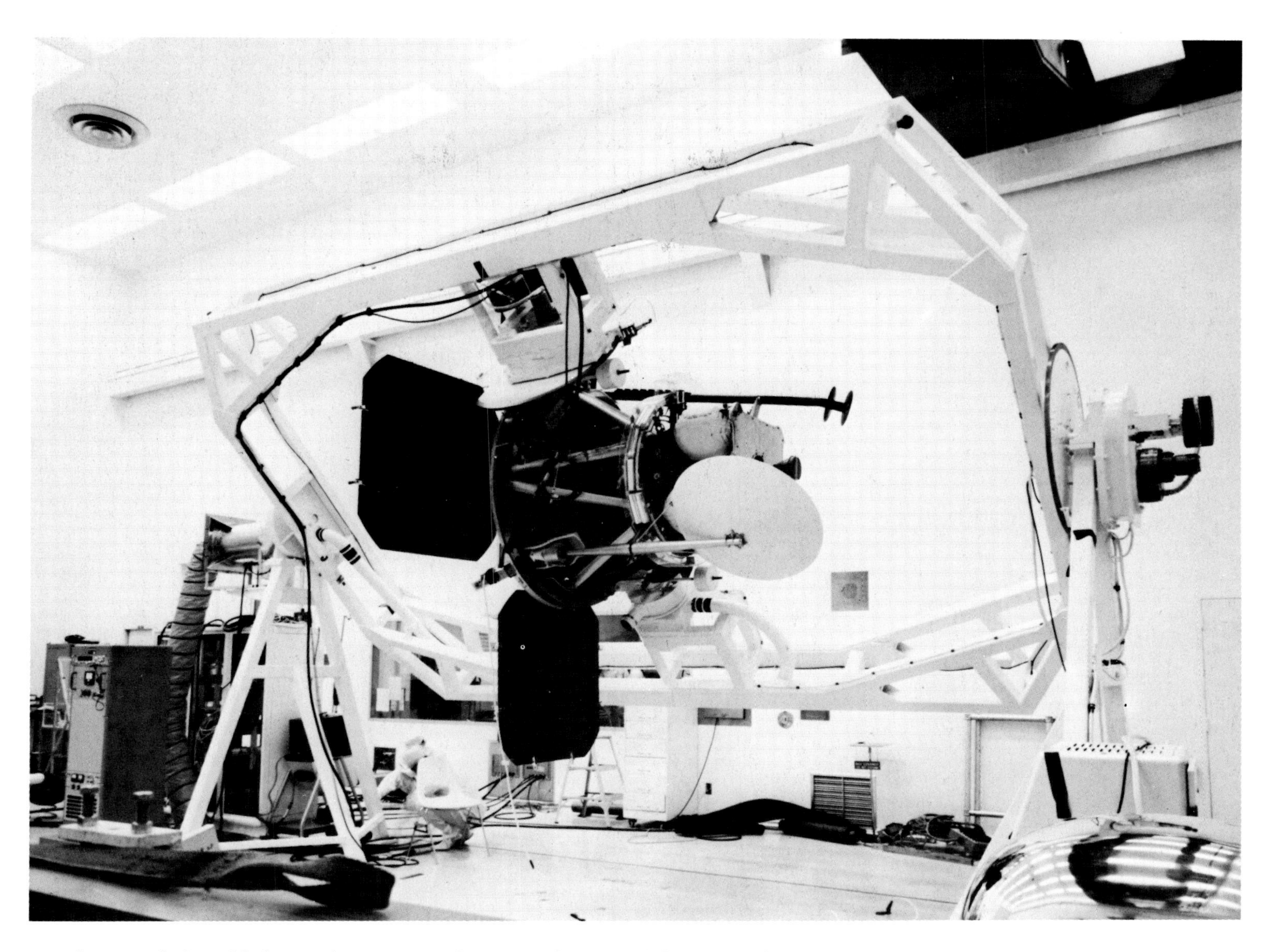

panels extended at 90-degree intervals. The velocity-control engine nozzle extended through the heat shield at the top of the cone structure.

Lunar Orbiter 1 surveyed nine primary and seven potential Apollo landing sites, including the Surveyor site.

Lunar Orbiter 2
Launch vehicle: Atlas Agena
Launch: 6 November 1966
Readout complete: 6 December 1966
Impact on the moon: 11 October 1967
Total weight: 861 pounds
Specifications and description: Same as Lunar Orbiter 1

Lunar Orbiter 2 surveyed 13 potential Apollo landing sites, photographed the wreckage of Ranger 8 on the lunar surface, responded to 2870 commands and performed 280 maneuvers.

Lunar Orbiter 3
Launch vehicle: Atlas Agena
Launch: 5 February 1967
Readout concluded: 23 February 1967
Impact on the moon: 9 October 1967
Total weight: 850 pounds
Specifications and description: Same as Lunar Orbiter 1

Lunar Orbiter 3 returned coverage of 600,000 square miles of the front and 250,000 square miles of the back side of the moon before the film advance broke down with 72 percent of the readout complete. Data was returned on 10 primary and 31 secondary Apollo landing sites.

Lunar Orbiter 4
Launch vehicle: Atlas Agena
Launch: 4 May 1967
Readout complete: 27 May 1967
Impact on the moon: 6 October 1967
Total weight: 860 pounds
Specifications and description: Same as Lunar Orbiter 1.

By the end of the fourth mission, the lunar orbiter program had surveyed 99 percent of the front and 80 percent of the back side of the moon.

Above: **Lunar Orbiter 4 mounted on a three-axis test stand with solar panels deployed.**
Below: **Scale models of the moon with the Lunar Orbiter.**

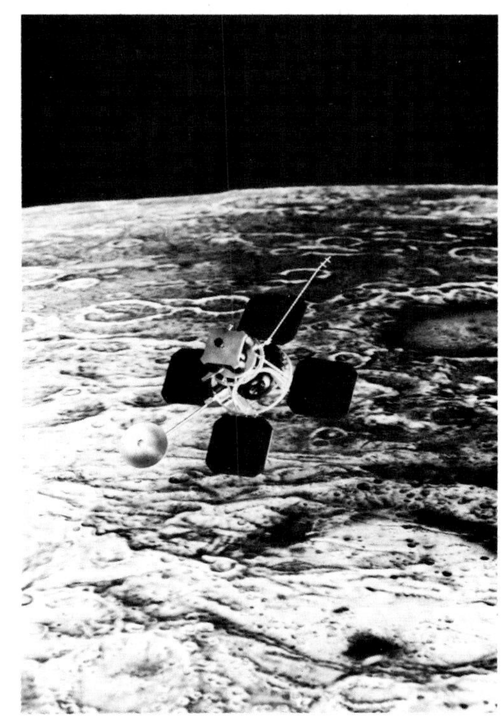

Leasat Spacecraft

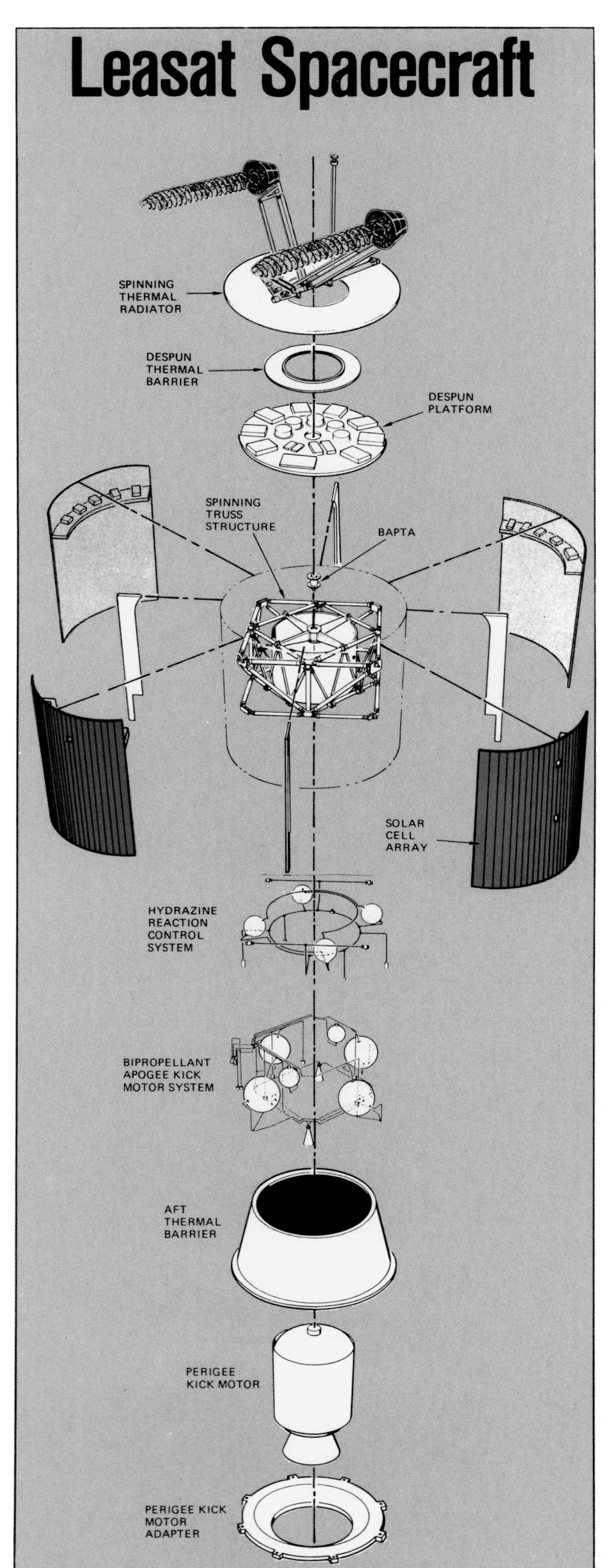

Lunar Orbiter

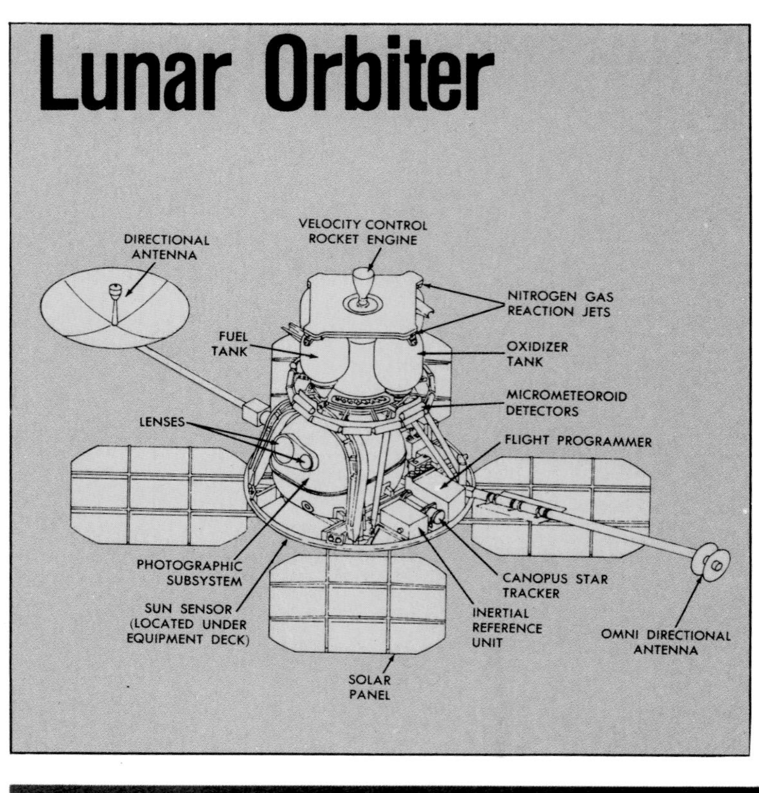

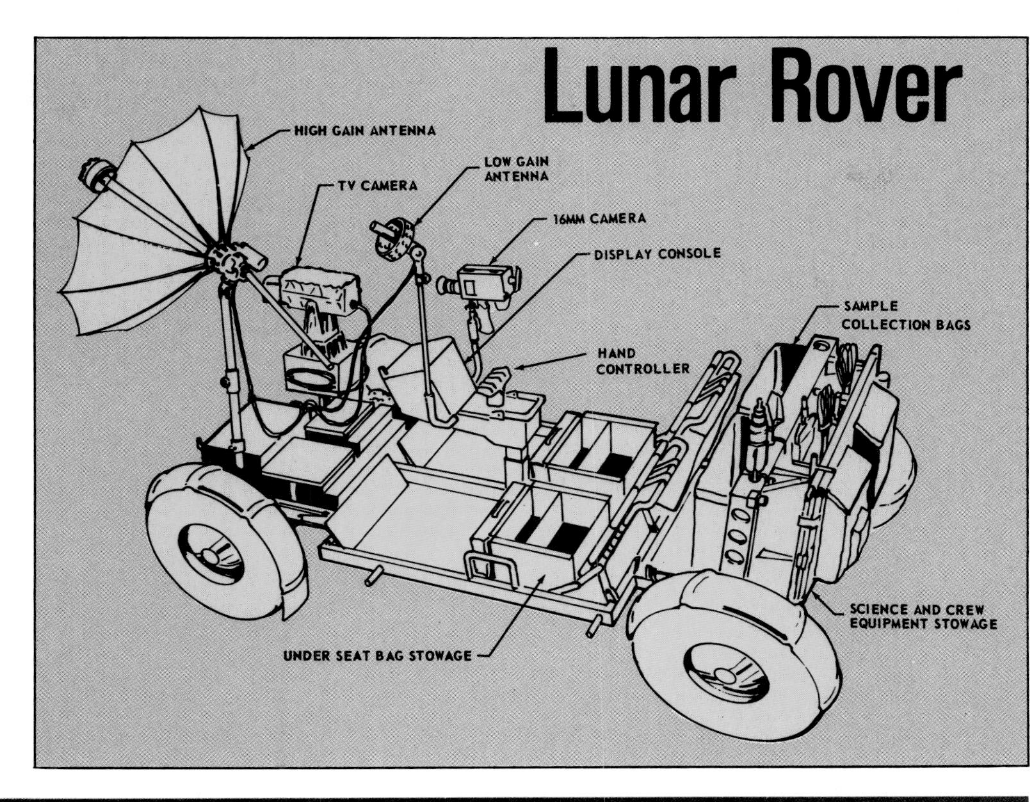

Lunar Orbiter 5
Launch vehicle: Atlas Agena
Launch: 1 August 1967
Readout completed: 31 August 1967
Impact on the moon: 31 January 1968
Total weight: 860 pounds
Specifications and description: Same as Lunar Orbiter 1

Lunar Orbiter 5 returned detailed coverage of five Apollo sites and 36 sites of scientific interest.

Lunar Roving Vehicle

Deployed: 31 July 1971 (from Apollo 15)
20 April 1972 (from Apollo 16)
11 December 1972 (from Apollo 17)
Vehicle weight: 462 pounds
Crossweight with maximum payload: 1542 pounds
Length: 137 inches
Width: 72 inches (to/from center of wheels)
Wheelbase: 90 inches
Power: Two 36-volt batteries (with one backup)
Design maximum speed: 8.7 mph
Actual top speed: 10.6 mph
Endurance: 78 hours (cumulative use)
Design range: 57 miles
Greatest distance covered: 17.5 miles (from Apollo 15)
16.6 miles (from Apollo 16)
22 miles (from Apollo 17)

The lunar rover (LRV) was unique among all US spacecraft in that it was not designed for space 'flight,' but to carry personnel over the surface of the moon. Developed by Boeing, the rover was essentially a small open automobile that could be driven by astronauts to carry them over greater distances than they could cover on foot. It was a compact electrically powered vehicle that could carry two astronauts or a combination of gear up to a total payload capacity of 1080 pounds (earth weight).

Three lunar rovers were used operationally and were carried to the moon aboard the Apollo J series spacecraft. They were carried folded into a small package in the pie-shaped confines of the Quad 1 bay of the Apollo landing module descent stage, that had been occupied on earlier missions by the erectable S-band antenna and laser reflector. When fully loaded the rovers could negotiate steplike obstacles 9.8 inches high and could cross crevasses 22.4 inches wide. They could climb and descend slopes as steep as 25

Lunar rover with Apollo spacecraft on the moon.

This lunar rover was taken to the moon by Apollo 15 in 1971.

degrees and stop and hold on slopes up to 30 degrees.

The rover's Boeing-built navigation system proved to be accurate and reliable. The astronauts reported that they did not hesitate to deviate from their planned route during their explorations, because the navigational system could easily return them to within sight of the Apollo landing module.

Each of the three lunar rovers were used successfully several times on each of their respective Apollo missions. They proved to be a valuable vehicle for exploration of the lunar surface and for hauling mineral samples back to the Apollo landing site for return to earth. Abandoned on the moon, the three rovers can be reclaimed for further use if subsequent lunar missions require them. The television system on the second rover was used to televise the launch of the Apollo 16 ascent stage from the lunar surface.

Magnetic Field Satellite

Launch vehicle: Scout G-1
Launch: 30 October 1979
Re-entry: 11 June 1980
Total weight: 400 pounds
Diameter: 30 inches (length from tip to tip of the deployed solar arrays was 134 inches)
Height: 63 inches

The Magnetic Field Satellite (Magsat) was made up of two modules: a base module, which used the same octagonal structure as the SAS, and an instrument module containing a science payload and instrumentation. Extending from the spacecraft were four solar array paddles that protruded from the main body in an 'X'-shaped configuration, and a 20-foot long scissors-type boom that deployed the Magsat's tow magnetometers to minimize magnetic interference from the spacecraft.

The Magsat was developed by the Applied Physics Laboratory of Johns Hopkins University and launched from the Western Space and Missile Center at Vandenberg AFB. It was the first spacecraft designed specifically to measure the near-earth magnetic field. It also returned data on geologic composition, rock formation, temperature and regional geologic structure of the earth.

Manned Orbiting Laboratory (MOL)

(See Appendix 1: Skylab Space Station)

Manned Spacecraft

The United States has operated four manned spacecraft programs: Mercury (1961), Gemini (1965), Apollo (1968) and the space shuttle (1981). Other related manned programs include the Lunar Roving Vehicle, designed for lunar surface travel, and the Skylab space station.

The space shuttle program was the first US spacecraft program to finally include women as crew members.
(See also Apollo, Gemini, Mercury, Lunar Roving Vehicle, Space Shuttle, Appendix 1: Skylab Space Station)

Manned Spacecraft, Code Names

In three of the US manned space programs, the individual spacecraft were given code names, much the same way that ships are given names. In some cases the spacecraft were named after famous ships. The manned Mercury spacecraft each carried the suffix '7' in deference to the seven US astronauts selected for the Mercury program. There were, however, only six named Mercury space flights. In two cases, names used by Apollo spacecraft appeared later as names given to space shuttle orbiters.

America, see Apollo 17 (CSM-114)
Antares, see Apollo 14 (LM-8)
Aquarius, see Apollo 13 (LM-7)
Atlantis, see Space Shuttle (OV-104)
Aurora 7, see Mercury MA-7
Casper, see Apollo 16 (CSM-113)
Challenger, see Apollo 17 (LM-12)
Challenger, see Space Shuttle (OV-99)
Charlie Brown, see Apollo 10 (CSM-106)
Columbia, see Apollo 11 (CSM-107)
Columbia, see Space Shuttle (OV-102)
Discovery, see Space Shuttle (OV-103)
Eagle, see Apollo 11 (LM-5)
Endeavor, see Apollo 15 (CSM-112)
Enterprise, see Space Shuttle (OV-101)
Faith 7, see Mercury MA-9
Falcon, see Apollo 15 (LM-10)
Freedom 7, see Mercury MR-3
Friendship 7, see Mercury MA-6
Gumdrop, see Apollo 9 (CSM-104)
Intrepid, see Apollo 12 (LM-6)
Kitty Hawk, see Apollo 14 (CSM-110)
Liberty Bell 7, see Mercury MR-4
Odyssey, see Apollo 13 (CSM-109)
Orion, see Apollo 16 (LM-11)
Sigma 7, see Mercury MA-8
Snoopy, see Apollo 10 (LM-4)
Spider, see Apollo 9 (LM-3)
Yankee Clipper, see Apollo 12 (CSM-108)

Mariner

The Mariners were a series of generally dissimilar spacecraft designed with the common purpose of conducting surveys of other planets in the solar system. The Mariner spacecraft were intended to explore and transmit data about other planets, principally Venus and Mars, from space nearby these planets without actually conducting a landing on the surface. With the exception of Mariners 5 and 10, which were intended for solo missions, the Mariner missions each involved two identical Mariner spacecraft launched toward the same planet roughly a month apart. In the case of Mariner 1/2, 3/4 and 8/9, however, one of the two spacecraft was lost on launch and these missions were actually flown solo. The Mariner 5 solo mission took advantage of the spacecraft that had been the backup for Mariner 4, and Mariner 10 was specifically conceived as a solo two-planet 'flyby' mission.

All of the Mariner missions except Mariner 8/9 and 10 were flyby missions, in which the spacecraft simply made a single pass near the planet to be surveyed. The Mariner 8/9 missions were intended to be orbital missions, and Mariner 9 was successfully placed into orbit around Mars. Mariner 10 made a single flyby of Venus and three flybys of Mercury.

An artist's concept of Mariner 1 passing Venus.

With the exception of the Mariner 10 spacecraft, which was designed and produced by Boeing Aerospace, all the Mariner spacecraft were developed by NASA's Jet Propulsion Laboratory.

Mariner 1
Launch vehicle: Atlas Agena B
Launch: 22 July 1962
Planet surveyed: Venus
First planetary encounter: None
Total weight: 447 pounds
Diameter: 5 feet
Height: 9 feet 11 inches
Shell composition: Sheet aluminum with gold plating as a thermal control
Shape: Tubular superstructure mounted on a hexagonal base

Mariner 1 was equipped with a microwave radiometer to determine the surface temperature and atmospheric details of

Venus; an infrared radiometer to assess cloud structure; a fluxgate magnetometer to measure magnetic fields; an iron chamber and three Geiger-Mueller tubes to measure the intensity and number of energetic particles; a cosmic dust detector and solar plasma detector. The spacecraft was destroyed on launch due to launch vehicle failure.

Mariner 2
Launch vehicle: Atlas Agena B
Launch: 27 August 1962
Planet surveyed: Venus
First planetary encounter: 14 December 1962
Description and systems: Same as Mariner 1

Mariner 2 was the first successful interplanetary probe.

Mariner 3
Launch vehicle: Atlas Agena D
Launch: 5 November 1964
Planet surveyed: Mars
First planetary encounter: None
Total weight: 575 pounds
Diameter: 22.6 feet
Height: 9.5 feet
Shell composition: Magnesium
Shape: Octagonal

Mariner 4 (*above left*) returned the first close-up photos of Mars. Mariner 6 (*above*) was almost twice as heavy as Mariner 5 (*left*).

A high-gain dish was mounted on top of the main structure of Mariner 3 along with low-gain antennae on top of an aluminum tube. Four solar panels extended from the top side of the octagon.

The spacecraft was equipped with a single camera to take and transmit photos to earth in digital form, a planetary experiment to study occultation of Martian atmospheric pressure and interplanetary experiments similar to those of Mariners 1 and 2.

The spacecraft shroud of Mariner 3 failed to jettison and communications with the spacecraft were lost.

Mariner 4
Launch vehicle: Atlas Agena D
Launch: 28 November 1964
Planet surveyed: Mars
First planetary encounter: 14 July 1965
Description and systems: Same as Mariner 3

Mariner 4 traveled 325 million miles to a distance of 6,118 miles from Mars, and returned 22 pictures as planned.

Mariner 5 (Venus 67)
Launch vehicle: Atlas Agena D
Launch: 14 June 1967
Planet surveyed: Venus
First planetary encounter: 19 October 1967
Total weight: 540 pounds
Diameter: 50 inches (main body)
Height: 20 inches
Shell composition: Magnesium
Shape: Octagonal main body

The Mariner 5 was a modified Mariner 4 backup. Four solar panels were deployed from the top, high-gain ellipse antennae were mounted on the superstructure on top of the body and an 88-inch tube on top supported the low-gain omnidirectional antennae and helium magnetometer. In flight the spacecraft was 18 feet across and the solar panels were 9 feet 6 inches across.

Planetary experiments on Mariner 5 included an S-band occultation based on spacecraft telemetry signals and ultraviolet photometry of upper atmosphere atomic hydrogen and oxygen radiation using three photomultiplier tubes. The spacecraft approached to within 2429 miles of Venus, and discovered 72 to 87 percent carbon dioxide atmosphere without oxygen.

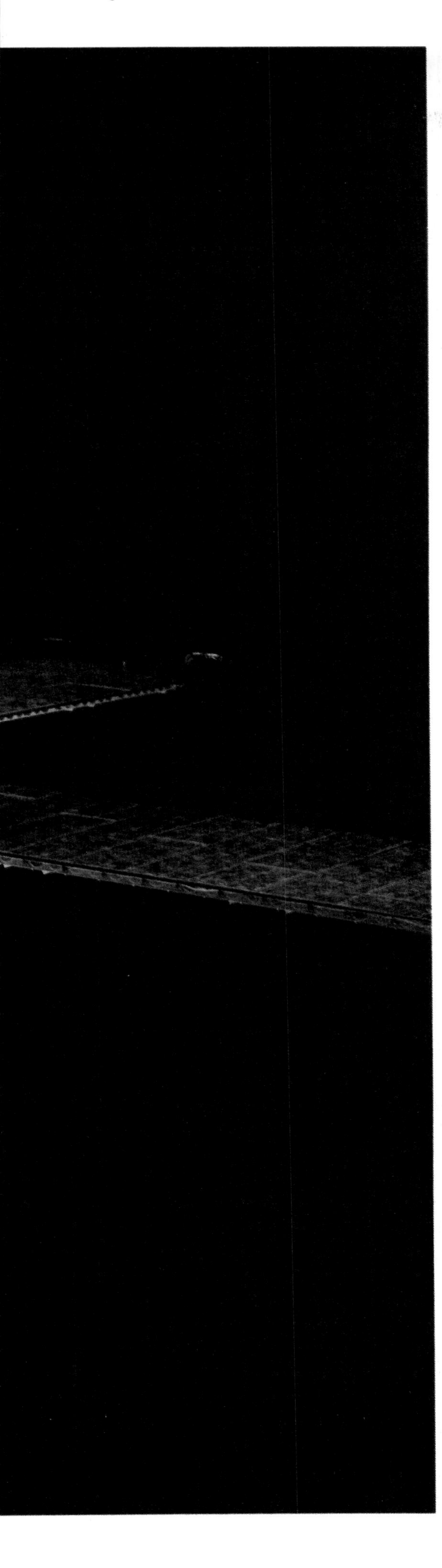

Mariner 6

Launch vehicle: Atlas Centaur
Launch: 25 February 1969
Planet surveyed: Mars
First planetary encounter: 31 July 1969
Total weight: 910 pounds

The base structure was a 37-pound octagonal forged-magnesium frame, 54.5 inches across diagonally and 18 inches deep, containing eight subsystem compartments that also provided structural strengthening. Four solar panels 84 inches long and 35.5 inches wide were attached to the top; each panel had 20.7 square feet of solar cells (83 square feet total for each spacecraft) that spanned 19 feet when deployed. Attitude-control jets were mounted at the panel tips. Low-gain omnidirectional antennae were mounted on top of a 4-inch-diameter aluminum tube, which served as a wave guide and extended 88 inches from the top of the base. A cone-shaped thermal control flux monitor was also mounted at the top of the mast. The total height was 11 feet from the top of the low-gain antennae masts to the bottom of the lower experiment-mount scan platform.

Mariner 6 carried systems that included: a wide-angle color television camera (similar to Mariner 4); a narrow-angle television camera (with 704 line resolution and equipped with a modified Schmidt Cassegrain telescope); an infrared radiometer (IRR) for thermal mapping; an ultraviolet spectrometer; an infrared spectrometer; a celestial mechanics experiment; tape recorders; a radio receiver (including S-band) and a transmitter.

The spacecraft passed within 2120 miles of Mars on 31 July, with a 54-minute span of experimentation including 41 minutes of photography in which a total of 75 photos were taken by the two cameras.

Mariner 7

Launch vehicle: Atlas Centaur
Launch: 27 March 1969
Planet surveyed: Mars
First planetary encounter: 5 August 1969
Description and systems: Same as Mariner 6

Mariner 7 passed within 2190 miles of Mars on 5 August, with a 72-hour span of experimentation in which two cameras took a total of 126 photos. As with Mariner 6, all other systems transmitted satisfactory results.

Mariner 8 (Mariner H)

Launch vehicle: Atlas Centaur

Above: **Mariner 7 with panels folded. *Left:* Mariner 9 had eight electronic compartments.**

Launch: 8 May 1971
Planet surveyed: Mars
First planetary encounter: None
Total weight: 2200 pounds (including a 40-pound framework)
Diameter: 54.5 inches
Height: 18 inches:
Solar panels: Four (84.5 inches by 35.5 inches)
Low-gain omnidirectional antennae: One (4 inches by 57 inches)
Shell composition: Forged magnesium framework
Shape: Octagonal
Power source: 14,742 photovoltaic cells on the solar panels generating 500 to 800 watts

The systems on Mariner 8 were the same as for Mariner 6 and 7. They also included an infrared interferometer spectrometer; data automation; and storage subsystems for sampling, digitizing, formatting, processing, storing and transmitting data from the scientific experiments.

Mariner 8 was destroyed on launch due to launch vehicle failure.

Mariner 9 (Mariner I)

Launch vehicle: Atlas Centaur
Launch: 30 May 1971
Planet surveyed: Mars
First planetary encounter: 13 November 1971
Description and systems: Same as Mariner 8

The Mariner 9 spacecraft was placed into orbit around Mars on 13 November,

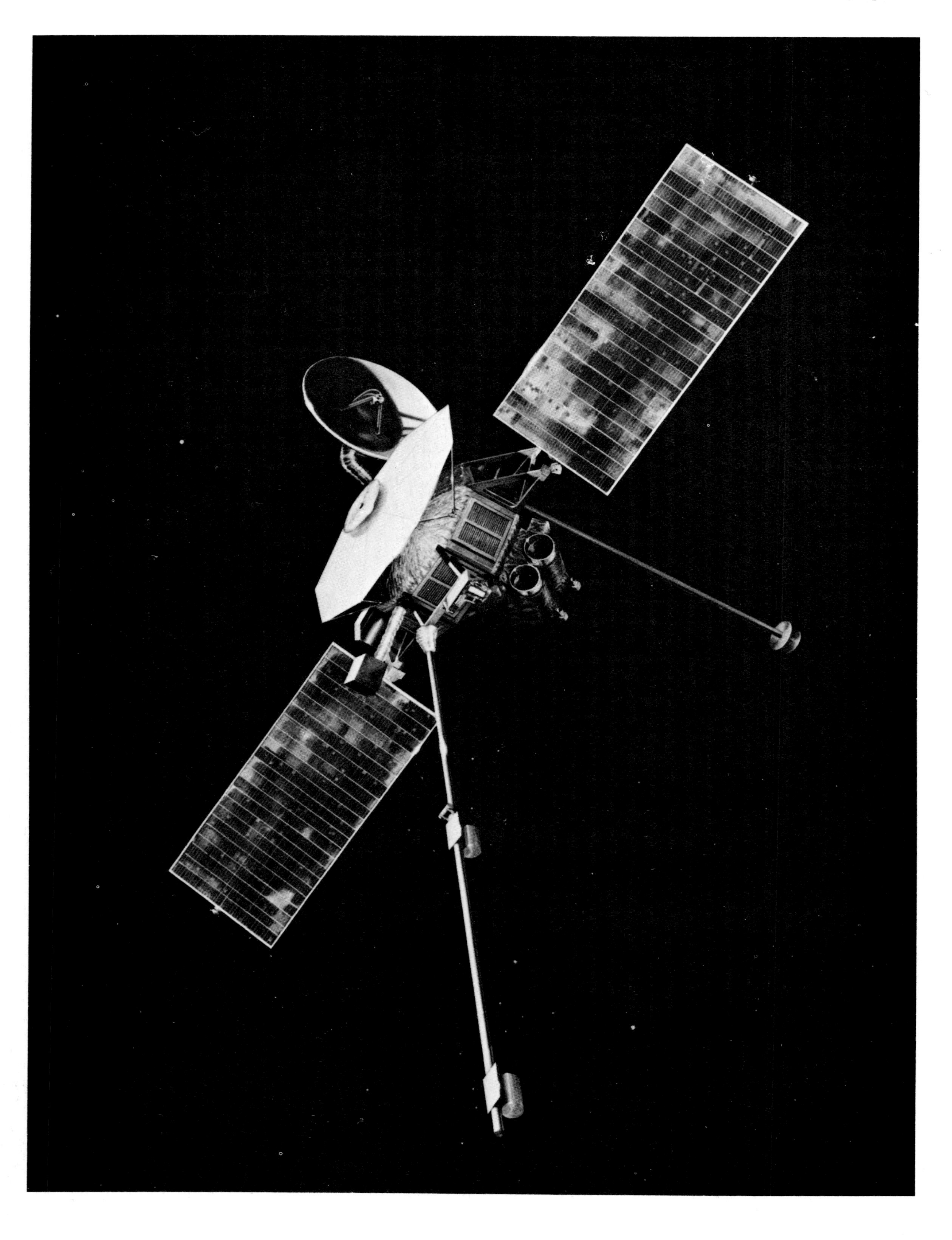

with a 90-day span of experimentation in which a total of 6876 photos of the Martian surface were returned. All of the scientific experiments were successfully completed.

Mariner 10
Launch vehicle: Atlas Centaur SLV-3C
Launch: 3 November 1973
Planets surveyed: Venus and Mercury
First planetary encounters: 5 February 1974 (Venus)
29 March 1974 (Mercury)
Total weight: 1108 pounds (including a 40-pound framework and a 172-pound experiment package)
Diameter: 54.5 inches
Height: 18 inches
Solar panels: Two (106 inches by 38 inches)
Low-gain omnidirectional antennae: Two (2.25 inches by 112 inches)
Motor-driven high-gain antenna diameter: 54 inches (honeycomb aluminum dish)
Shell compositon: Forged magnesium framework
Shape: Octagonal

The systems included in Mariner 10 were similar to those of Mariner 8 and 9.

The spacecraft passed Venus on 5 February 1974 as scheduled and Mercury for the first time on 29 March as scheduled. A second flyby of Mercury, on 21 September 1974, was a day earlier than planned. The spacecraft returned 8000 photographs as well as other scientific data, and was shut down and placed in permanent solar orbit on 24 March 1975, after a third unscheduled flyby of Mercury.

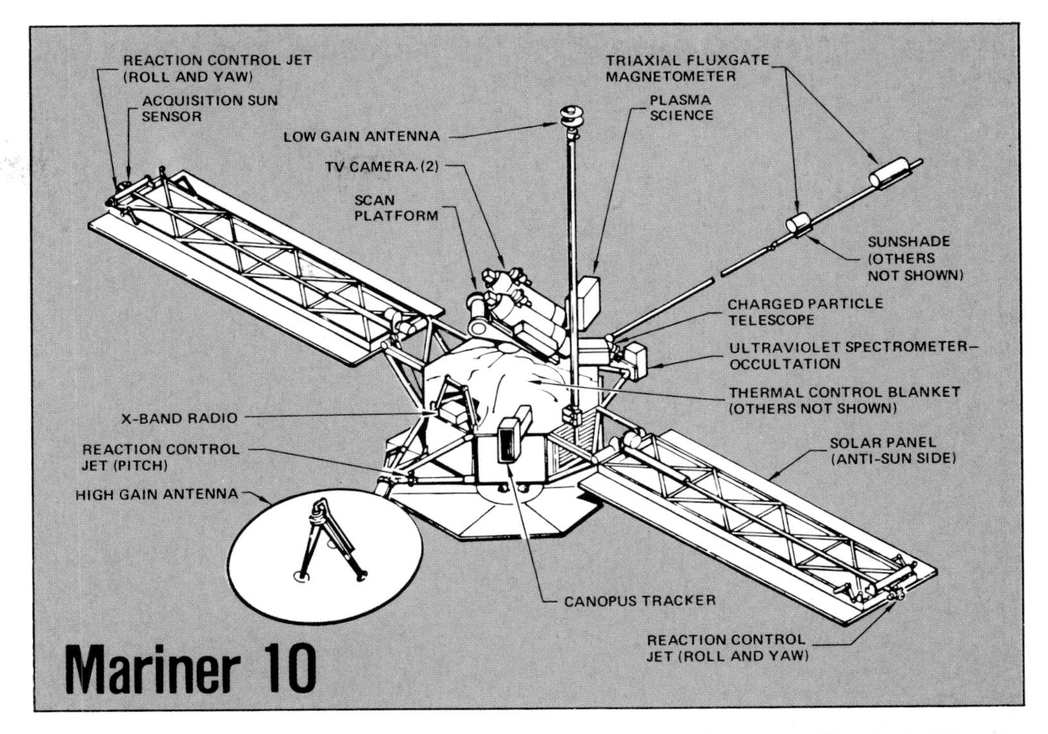

Left: Mariner 10, NASA's first dual-planet mission and the first to explore Mercury (*below*).

Above: Mariner 8 carried two television cameras and experiments in the bottom scan platform.

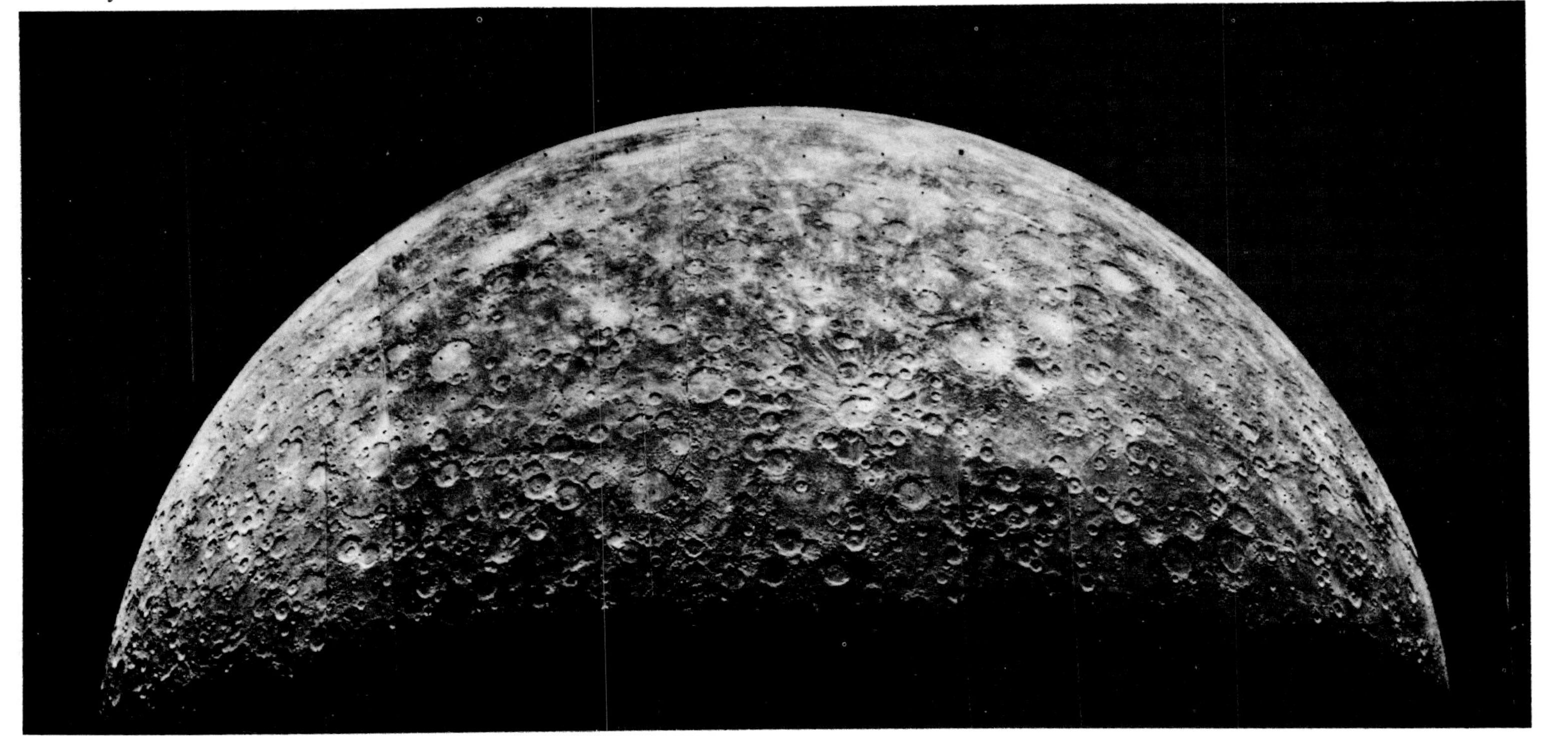

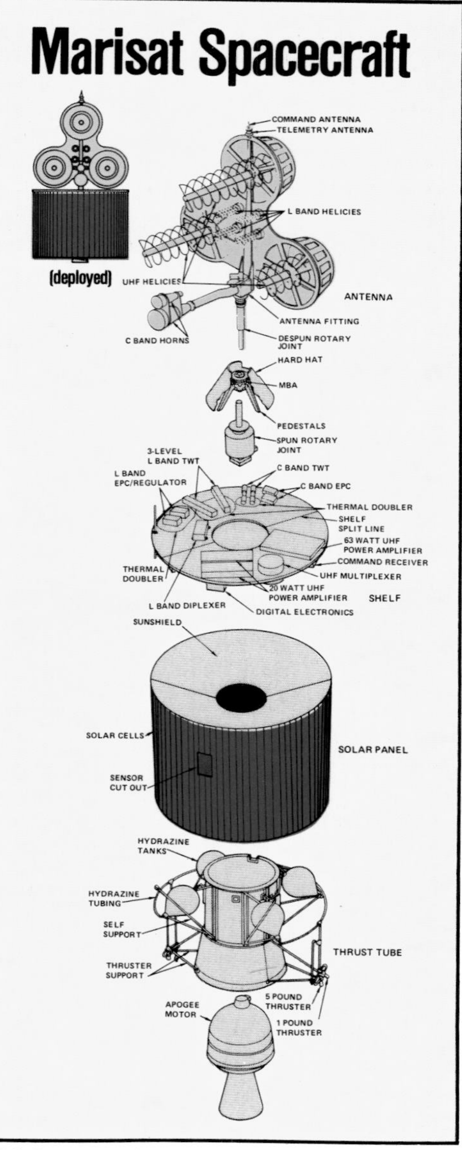

Marisat 1 was the first satellite of COMSAT's worldwide maritime communications system.

Marisat

Launch vehicle: Delta Straight Eight
Launch: 19 February 1976 (Marisat 1)
9 June 1976 (Marisat 2)
14 October 1976 (Marisat 3)
Total weight: 1445 pounds
Diameter: 85 inches
Height: 63 inches (148 inches with antennae)
Shape: Cylindrical
Frequencies: 240-400 MHz (UHF)
1.4-1.6 GHz (L-Band)
4 and 6 GHz (C-Band)
Channels: 44 teletype or 14 voice (UHF)
14-22 teletype or voice (Land C-Band) at 18.8 to 29.5 dBW

The Marisat spacecraft were maritime communcations satellites designed and produced by Hughes Aircraft and launched by NASA on a reimbursable basis for a consortium of four companies headed by COMSAT General Corporation (formerly the Communications Satellite Corporation). The consortium included COMSAT (86.3 percent), RCA Global Communications (8 percent), Western Union International (3.4 percent) and ITT World Communications (2.3 percent). The primary purpose of the spacecraft was to provide common-carrier service to ships of all nations. Marisat 1 (A) was placed in a geosynchronous equitorial orbit over the Atlantic, Marisat 2 (B) was placed in a similar orbit over the Pacific and Marisat 3 (C) was parked above the Indian Ocean. Both the US Maritime Administration and the US Navy were major users of the Marisat system, although the Navy was the only user of Marisat 3 when it was first placed on station. The spacecraft were powered by 7000 solar cells delivering 330 watts.

Mercury

Spacecraft height: 9 feet (Mercury Atlas 3-4, Mercury Redstone 3-4)
9 feet 6 inches (Mercury Atlas 5-9)
Escape tower height: 16 feet
Base diameter: 6 feet

Right: **Mercury MR-3 (*Freedom 7*) in May 1961 carried Alan Shepard on what was to be the first US manned spaceflight.**

Weight: Variable (2000 to 3000 pounds)
Material: 0.01-inch titanium skin with an ablative fiberglass heat shield

The objective of the Mercury program was to place an American astronaut in outer space for the first time. The program was begun in November 1958, and NASA issued a contract to the McDonnell Aircraft Company on 13 February 1959 for 12 one-crew-member spacecraft. The total was later increased to 20 flight-rated capsules. Of these, six of the last seven were the manned missions, which took place between May 1961 and May 1963. Seven astronauts were selected, and each of them except Donald Slayton completed one Mercury flight (see also Apollo-Soyuz Test Project). Despite a 43 percent failure rate in the 14 test flights leading up to the first manned flight, all of the manned Mercury missions were completed successfully.

The Mercury flights were denoted under five designation systems. The first three of these designation types were by launch vehicle: Mercury Atlas (MA), Mercury (Big and Little) Joe, Mercury Redstone (MR) and Mercury Scout (MS). Those eight spacecraft that carried a living organism were given a Mercury sequence number *in addition* to their launch vehicle designator. Finally, the six spacecraft that carried men were *also* given a code name containing the numeral 7, which was not a sequence numeral but a reference to the seven men in the first group of astronauts. For the sake of clarity all the flights of the Mercury program are listed in chronological order, together with their various designation numbers where applicable.

Mercury Big Joe
Launch vehicle: Atlas (Big Joe)
Launch and recovery: 9 September 1959
Crew: None

Mercury Big Joe performed a successful unmanned suborbital capsule test.

Mercury Little Joe 1
Launch vehicle: Little Joe
Launch and re-entry: 4 October 1959
Crew: None

Mercury Little Joe 1 completed a successful unmanned suborbital capsule test.

Mercury Little Joe 2
Launch vehicle: Little Joe

Above: The first Mercury test with Atlas Big Joe was on 9 September 1959.

Above: A flight-ready Mercury 6. Below: Mercury MR-3 is mated to a Redstone booster.

Launch and re-entry: 4 November 1959
Crew: None

Mercury Little Joe 2 failed as a result of launch vehicle failure.

Mercury Little Joe 3
Launch vehicle: Little Joe
Launch and re-entry: 4 December 1959
Crew: Sam (Rhesus monkey)

Mercury Little Joe 3 performed a successful suborbital capsule test as well as biomedical and escape-system tests.

Mercury Little Joe 4
Launch vehicle: Little Joe
Launch and recovery: 21 January 1960
Crew: Miss Sam (Rhesus monkey)

Mercury Little Joe 4 also completed a successful suborbital capsule test with biomedical and escape-system tests.

Mercury MA-1
Launch vehicle: Atlas
Launch and re-entry: 29 July 1960
Crew: None

Mercury MA-1 resulted in launch vehicle failure.

Mercury Little Joe 5
Launch vehicle: Little Joe
Launch and re-entry: 8 November 1960
Crew: None

The escape system test of Mercury Little Joe 5 failed when the escape rocket fired prematurely.

Mercury MR-1
Launch vehicle: Redstone
Launch and re-entry: 19 December 1960
Crew: None

Mercury MR-1 completed a successful 235-mile unmanned suborbital test flight, the first Mercury capsule to actually go into outer space.

Mercury MR-2 (Mercury 2)
Launch vehicle: Redstone
Launch and re-entry: 31 January 1961
Crew: Ham (Chimpanzee)

After a successful 16-minute suborbital test flight, Mercury MR-2 was the first American flight into space with a living organism aboard.

Mercury MA-2
Launch vehicle: Atlas
Launch and recovery: 21 February 1961
Crew: None

Mercury MA-2 resulted in a successful 1425-mile unmanned suborbital test flight.

Mercury Little Joe 5A
Launch vehicle: Little Joe
Launch and recovery: 18 March 1961
Crew: None

The escape system test of Mercury Little Joe 5A failed when the escape rocket fired prematurely.

Mercury MR-BD (Booster Development)
Launch vehicle: Redstone
Launch and re-entry: 24 March 1961
Crew: No spacecraft attached

This launch was simply a vehicle test without the Mercury capsule.

Mercury MA-3
Launch vehicle: Atlas
Launch and recovery: 25 April 1961
Crew: None

Mercury MA-3 experienced launch vehicle failure but the capsule was successfully recovered.

Mercury Little Joe 5B
Launch vehicle: Little Joe
Launch and recovery: 28 April 1961
Crew: None

The escape system test similar to the one undertaken by Little Joe 5 and 5A was unsuccessful due to launch vehicle failure.

Mercury MR-3 (Mercury 3)
Freedom 7
Launch vehicle: Redstone
Launch and recovery: 5 May 1961
Crew: Alan Shepard

The first American manned spaceflight, in *Freedom 7,* lasted 15 minutes, and the capsule was recovered in the Atlantic after a successful suborbital flight.

Mercury MR-4 (Mercury 4)
Liberty Bell 7
Launch vehicle: Redstone
Launch and recovery: 21 July 1961
Crew: Virgil Grissom

The second American manned spaceflight aboard *Liberty Bell 7* lasted 16 minutes. Grissom was recovered in the Atlantic (the capsule sank) after a successful suborbital flight.

The Mercury MA-6 Atlas was launched after a month of delays and last-minute problems.

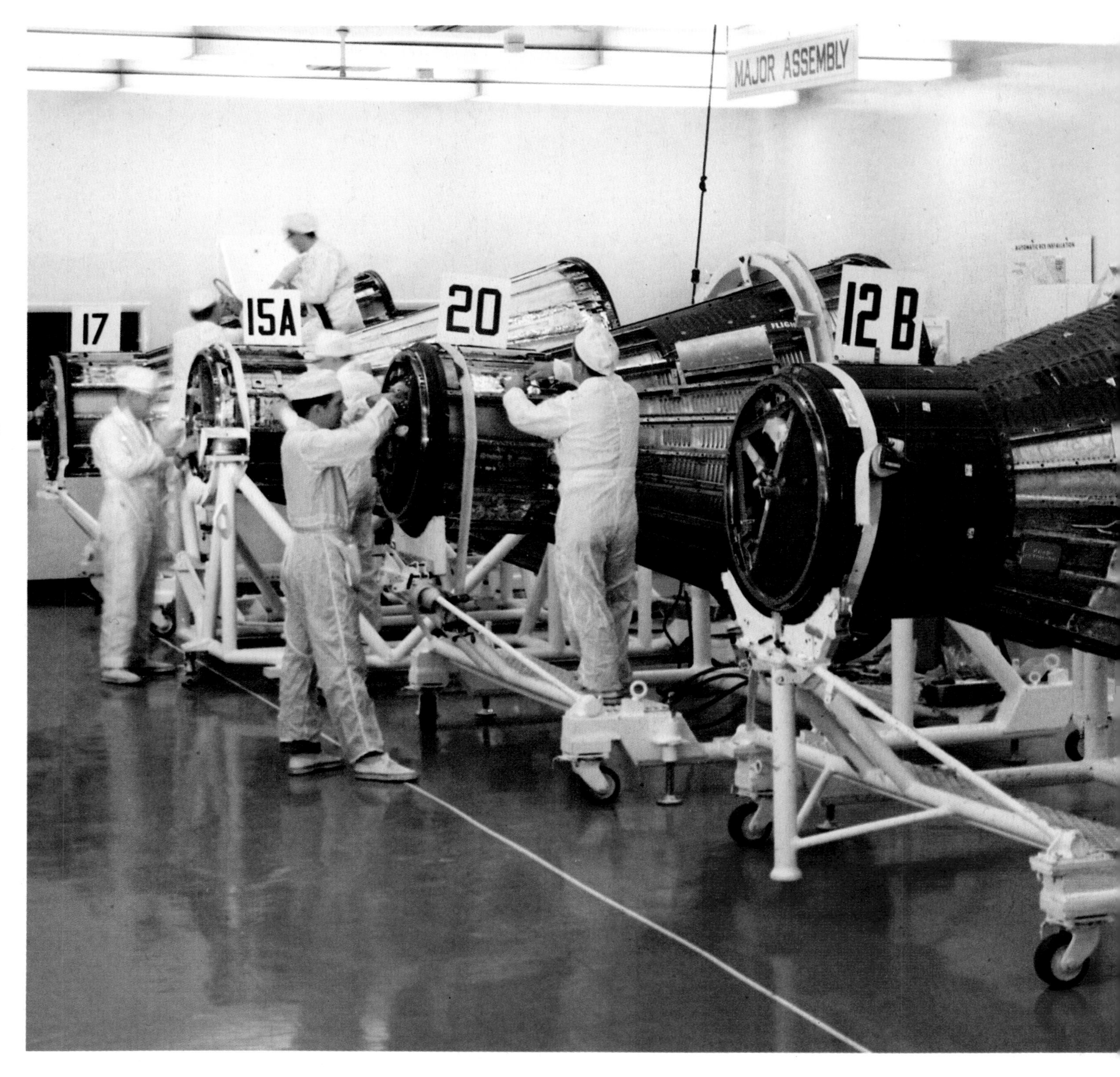

Mercury MA-4
Launch vehicle: Atlas D
Launch and recovery: 13 September 1961
Crew: None

The Mercury MA-4 performed the first successful orbital test flight of an unmanned Mercury capsule.

Mercury MS-1
Launch vehicle: Blue Scout
Launch and re-entry: 1 November 1961
(US Air Force launch 609 A)
Crew: None

The test of the Mercury orbital tracking network was not completed due to launch vehicle failure.

Mercury MA-5 (Mercury 5)
Launch vehicle: Atlas D
Launch and recovery: 29 November 1961
Crew: Enos (Chimpanzee)

Mercury 5 carried a living organism on the first orbital flight of a Mercury capsule and was also the last unmanned Mercury flight. Mercury 5 completed two orbits of a planned three, but was returned to earth early as a result of simple malfunctions with the roll reaction jet control and environmental control system that could have been easily corrected by a human astronaut.

Mercury MA-6 (Mercury 6)
Friendship 7
Launch vehicle: Atlas D
Launch and recovery: 20 February 1962
Crew: John Glenn

The first American manned orbital spaceflight completed three orbits in 4

Above: **The Mercury capsule assembly area at McDonnell Douglas, St Louis in March 1962.**

hours 55 minutes, and was successfully recovered.

Mercury MA-7
Aurora 7
Launch vehicle: Atlas D
Launch and recovery: 24 May 1962
Crew: Scott Carpenter

The second American manned orbital spaceflight completed three orbits in 4 hours 56 minutes, and was successfully recovered.

Above and below: **The *Liberty Bell* (MR-4) is positioned atop a Redstone launch vehicle.**

Mercury MA-8 (Mercury 8)
Sigma 7
Launch vehicle: Atlas D
Launch and recovery: 3 October 1962
Crew: Walter Schirra

Sigma 7 achieved a manned orbital spaceflight of 9 hours 13 minutes, more than double the duration of the earlier Mercury orbital missions, and was successfully recovered after six orbits.

Mercury MA-9 (Mercury 9)
Faith 7
Launch vehicle: Atlas D
Launch: 15 May 1963
Recovery: 16 May 1963
Crew: Gordon Cooper

The longest and last of the Mercury spaceflights completed 22 orbits in 34 hours 20 minutes, and made it possible to evaluate the performance of an astronaut during a longer duration in space.

Mercury MA-10 (Mercury 10)
The scheduled flight of Mercury 10 was canceled in order to proceed with flights of the Gemini two-crew-member spacecraft program.

Meteoroid Satellite

(See Explorers S-55, 13, 16, 23, 46)

Meteorological Satellite Programs

(See DMSP, ESSA, GOES, ITOS, Landsat, Nimbus, NOAA, TIROS)

MIDAS

Launch vehicle: Atlas Agena D
Launch: 26 February 1960 (Midas 1)
24 May 1960 (Midas 2)
12 July 1961 (Midas 3)
21 October 1961 (Midas 4)
9 May 1963 (Midas 6)
5 October 1966 (Midas 9)

An outdated US military system of early-warning satellites, MIDAS (Missile Defense Alarm System) preceded IMEWS in the role of detecting missile launches, including possible enemy ICBM launches. MIDAS 6 and 9 are considered to be part of the Ballistic Missile Early Warning System (BMEWS). Most of the data concerning MIDAS remains classified.

MOL

(See Appendix 1: Skylab Space Station)

NATO

The NATO spacecraft (known as NATOSAT before 1971) were designed to provide a communications link between NATO headquarters near Brussels, the capitals of NATO member nations, and NATO land and sea commands. These geosynchronous-orbit spacecraft were developed in the US by Philco-Ford (later Ford Aerospace), launched by NASA, and are operated under the command and control of the US Air Force Satellite control facility near San Francisco.

NATOSAT 1 (NATO-A)
Launch vehicle: Thrust Augmented Delta (TAD)
Launch: 20 March 1970
Total weight: 535 pounds at launch
285 pounds in orbit
Diameter: 54 inches
Height: 63 inches
Shape: Cylindrical

NATOSAT 1 was equipped with two transponders, each with two channels to receive, frequency translate, amplify and retransmit traffic originated at ground stations, with provisions for large high-

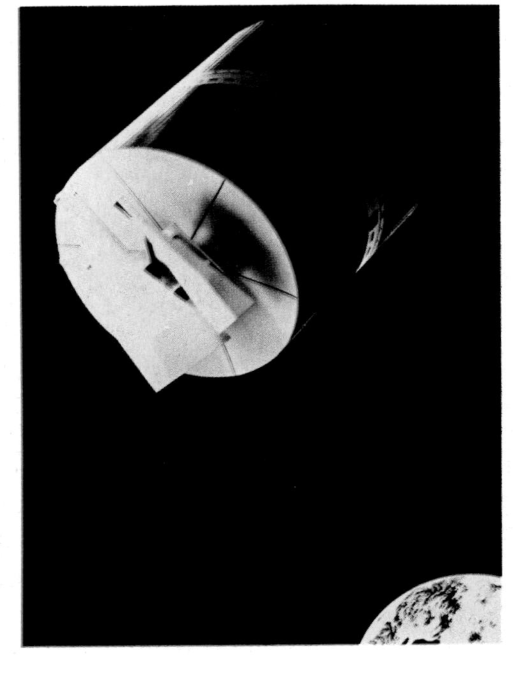

Left: NATO satellites relay communications from 22,300 miles above the Atlantic.

power fixed-ground stations as well as small mobile stations.

NATO 2 (NATO-B)
Launch vehicle: Thrust Augmented Delta (TAD)
Launch: 2 February 1971
Description and systems: Same as NATOSAT 1/NATO-A

NATO 3A
Launch vehicle: Delta 2914
Launch: 22 April 1976
Total weight: 1528 pounds at launch
830 pounds in orbit
Diameter: 86 inches
Height: 122 inches (including antennae)
Shape: Drum

NATO 3A had three 20-watt traveling wave tubes receiving, frequency translating and retransmitting voice, telegraph, facsimile and wide-band digital data in the 7 to 8 GHz frequency range.

NATO 3B
Launch vehicle: Delta 2914
Launch: 27 January 1977
Total weight: 1528 pounds at launch
830 pounds in orbit
Diameter: 86 inches
Height: 88 inches
Shape: Drum

The NATO 3B has an overall length of 122 inches, including antennae, and carries the same sytsems as NATO 3A. Its life expectancy is seven years.

NATO 3C
Launch vehicle: Delta 2914
Launch: 19 November 1978
Total weight: 1587 pounds at launch
829 pounds in orbit
Diameter: 7.2 feet
Height: 10 feet (with antennae extended)
Shape: Drum
Systems: Same as NATO 3A

NATO 3D
Launch vehicle: Delta 2914
Launch: 13 November 1984
Description and systems: Same as NATO 3B

Navigation Technology Satellites (NTS)

(See NAVSTAR, NOVA, Timation, Transit, Triad)

Navsat (Naval Navigation Satellite)

(See Transit)

NAVSTAR

Launch vehicle: Atlas E/F
First launch: 22 February 1978
Total weight: 1636 pounds at launch
1016 pounds in orbit
Width: 210 inches (with solar panels extended)

The NAVSTAR (Navigation System Using Time and Ranging) satellites developed by Rockwell International are the spacecraft element of the US Air Force/US Navy Global Positioning System (GPS). It is a successor to the US Navy Timation, Transit and Triad navigational satellite systems. The system is designed to provide pinpoint navigational accuracy to land, sea and air forces in the field. The GPS is designed to function with a galaxy of 18 NAVSTAR spacecraft, nine of which had been launched by July 1984. The first Block 2 spacecraft will be in orbit in 1986, and the complete system will be in place by 1988. The spacecraft are three-axis stabilized and placed in several overlapping planes circling the globe every 12 hours. NAVSTARs provide accurate three-dimensional position fixes within 52 feet with velocity to within 4 inches per second.

The ground portions of the GPS include a master control station at US Air Force Space Command (SPACECOM) Headquarters in Colorado and monitoring stations in the Indian Ocean (Diego

Garcia), the Atlantic Ocean (Ascension Island) and the Pacific (Hawaii, Guam and the Philippines). Small user packs are also distributed to small units of all branches of the US armed services, including 7291 USAF aircraft, and as many as 4700 US Navy ships. Spacecraft, such as Landsat, also carry GPS receivers.

NAVSTAR carries an atomic clock to provide timing data accurate to within one millionth of a second. An L-band transmitter provides positioning data, and has a pseudorandom noise code incorporated into the L-band transmissions that requires a match by the receiving station to prevent an unauthorized user from getting into the system.

Above: Military applications of NAVSTAR include precision mapping.

Below: An artist's concept of the multiservice international NAVSTAR constellation.

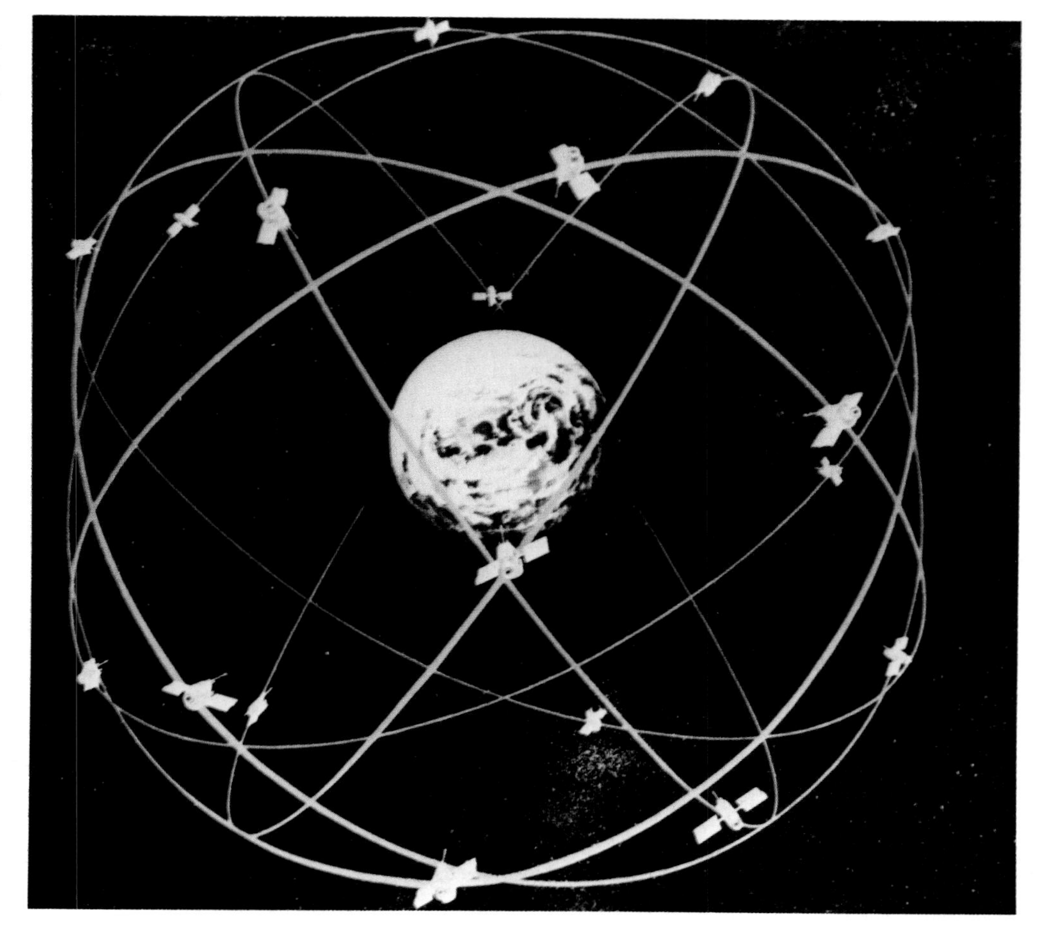

NERVA
(See RFD)

Nimbus

Originally conceived as a weather forecasting program (*nimbus* is Latin for rain cloud), the later satellites in this series became sufficiently sophisticated to provide data for a wider range of scientific programs including agriculture, carto-

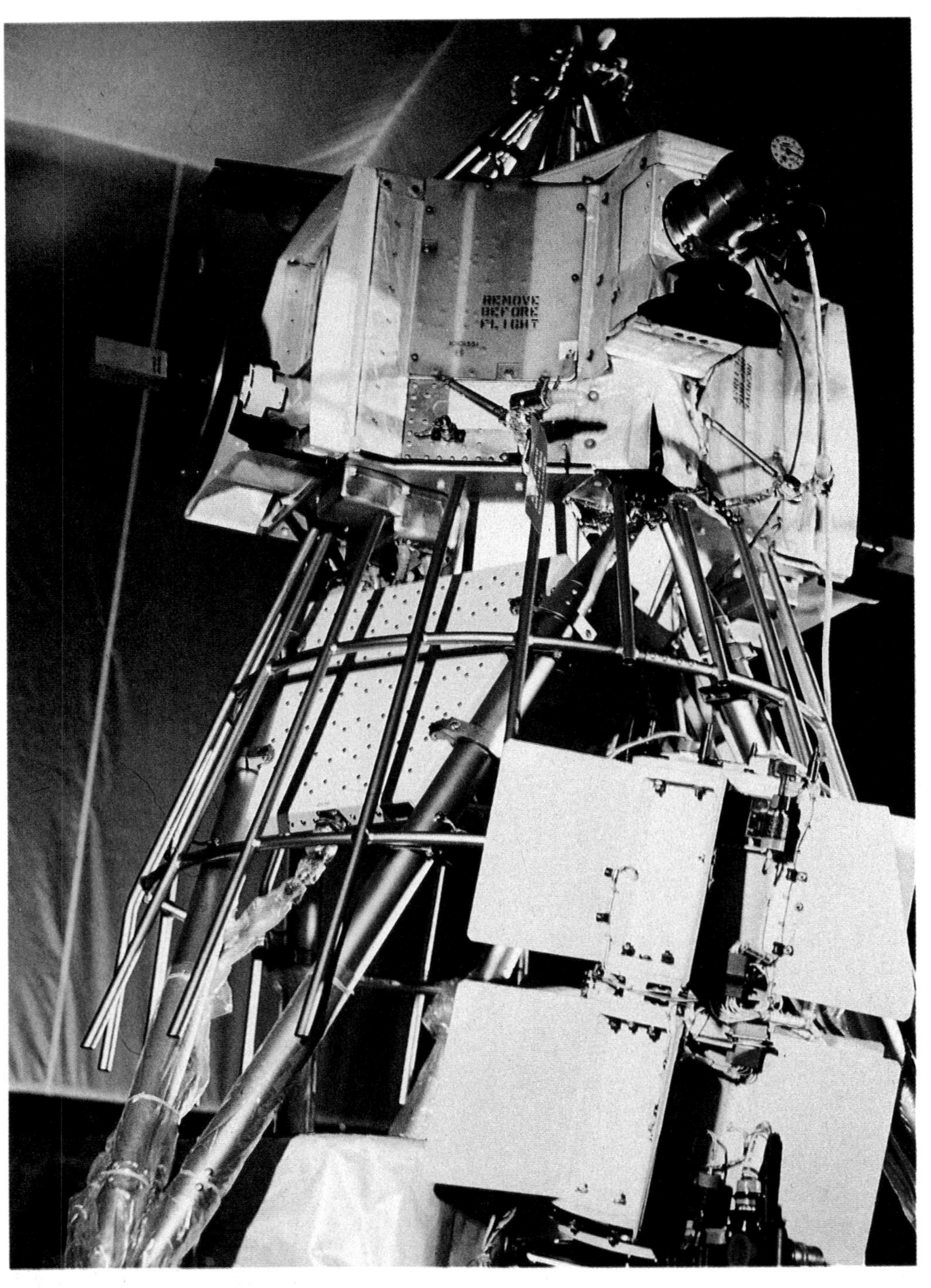

The ill-fated Nimbus B weather satellite.

graphy, geography, geology, hydrology and oceanography. The Nimbus spacecraft have been especially important in providing data in arctic and antarctic ice flows. The more recent Nimbus operations have been similar to those of Landsat. Developed by General Electric, these spacecraft are launched from the Western Space and Missile Center at Vandenberg AFB and managed by NASA's Goddard Space Flight Center.
(See also ESSA, ITOS, Landsat, NOAA, TIROS)

Nimbus 1
Launch vehicle: Thor Agena B
Launch: 28 August 1964
Re-entry: 16 May 1974
Total weight: 830 pounds
Diameter: 57 inches
Height: 118 inches
Width: 132 inches (across solar panels)

A magnesium truss connected two large solar panels of the hexagonal main section to the lower section ring sensor. The solar panels were attached to the sides of the Nimbus like butterfly wings and provided 200 watts of power. The spacecraft systems included an automatic picture transmission (APT) for delayed or direct readout, advanced vidicon camera system (AVCS) for daytime observations and high-resolution infrared radiometer (HRIR) for nighttime and cloud-cover observations.

Nimbus 1 provided complete global cloud-cover photography every 24 hours and spotted five hurricanes and two typhoons. A total of 27,000 photos were returned before the solar panel drive failed on 23 September 1964.

Nimbus 2
Launch vehicle: Thor Agena B
Launch: 15 May 1966
Last transmission: 17 January 1966
Total weight: 912 pounds
Specifications and description: Same as Nimbus 1
Systems: Same as Nimbus 1, plus a five-channel medium-resolution infrared radiometer (MRIR)

Nimbus 2 provided extensive coverage of the hurricane breeding area of the Atlantic, and exceeded its intended six-month lifespan by two months.

Nimbus B
Launch vehicle: Thor AD/Agena D
Launch and recovery: 18 May 1968
Total weight: 1260 pounds
Specifications and description: Same as Nimbus 1
Systems: Same as Nimbus 2 plus an infrared inferometer spectrometer (IRIS) to measure air-pollution dispersion; a satellite infrared spectrometer (SIRS) to measure water vapor and ozone dispersion in the atmosphere; interrogation, recording and location system (IRLS) to locate and identify transmissions from buoys, balloons and other small weather stations; and SNAP-19 experimental radioactive two-unit isotopic plutonium 238-powered electrical generators (recovered after the crash and reused on Nimbus 3)

Nimbus B crashed into the Pacific Ocean off California after the launch vehicle failed.

Nimbus 3
Launch vehicle: Thor Augmented Delta/Agena D
Launch: 14 April 1969
Re-entry: 29 December 1971
Total weight: 1269 pounds (heaviest meteorological satellite to date)
Specifications and description: Same as Nimbus 1
Systems: Same as Nimbus B plus image dissector camera (IDC) that replaced the AVCS, and a monitor of ultraviolet solar energy (MUSE)

All systems performed perfectly on Nimbus 3, which replaced Nimbus B. The

IRLS was used experimentally to monitor and precisely locate a tiny ground transmitter attached to an elk in Yellowstone National Park.

Nimbus 4
Launch vehicle: Thor AD/Agena D
Launch: 8 April 1970
Total weight: 1366 pounds
Specifications and description: Same as Nimbus 1
Systems: Same as Nimbus 3 plus a backscatter ultraviolet spectrometer (BUS); a filter wedge spectrometer (FWS); a selective chopper radiometer (SCR); and a temperature, humidity and infrared radiometer (THIR)

The results of this mission were successful.

Nimbus 5 (Nimbus E)
Launch vehicle: Delta 900
Launch: 11 December 1972
Total weight: 1580 pounds
Specifications and description: Same as Nimbus 1
Systems: Same as Nimbus 4, except it lacked the SNAP-19 power generators (see Nimbus B), and had an electrically scanning microwave radiometer (ESMR), an infrared temperature profile radiometer (ITPR), a surface composition mapping radiometer (SCMR) and the Nimbus E microwave spectrometer

The Nimbus 5 completed a successful mission.

Nimbus 6 (Nimbus F)
Launch vehicle: Delta 2910

Above left: **Nimbus 6 carried improved systems.**
Above: **The Nimbus 5 applications observatory.**

Launch: 12 June 1975
Total weight: 1823 pounds
Specifications and description: Same as Nimbus 1
Systems: Same as Nimbus 5

Nimbus 6 completed a successful operation and obtained data that helped in the development of the ITOS spacecraft.

Nimbus 7 (Nimbus G)
Launch vehicle: Delta 2910
Launch: 24 October 1978
Total weight: 2171 pounds
Specifications and description: Same as Nimbus 1
Systems: Same as Nimbus 5 plus a limb-scanning infrared radiometer (LSIRR), a stratospheric and mesopheric sounder (SMS), a stratospheric aerosol measurement device (SAMD), an earth radiation budget system (ERBS), and a coastal zone color scanner (CZCS)

The successful operation of the Nimbus 7 spacecraft systems concentrated on oceanography and pollution monitoring. It also released the Project Cameo (chemically active material into orbit) package designed to study the North Polar cap and the auroral belt.

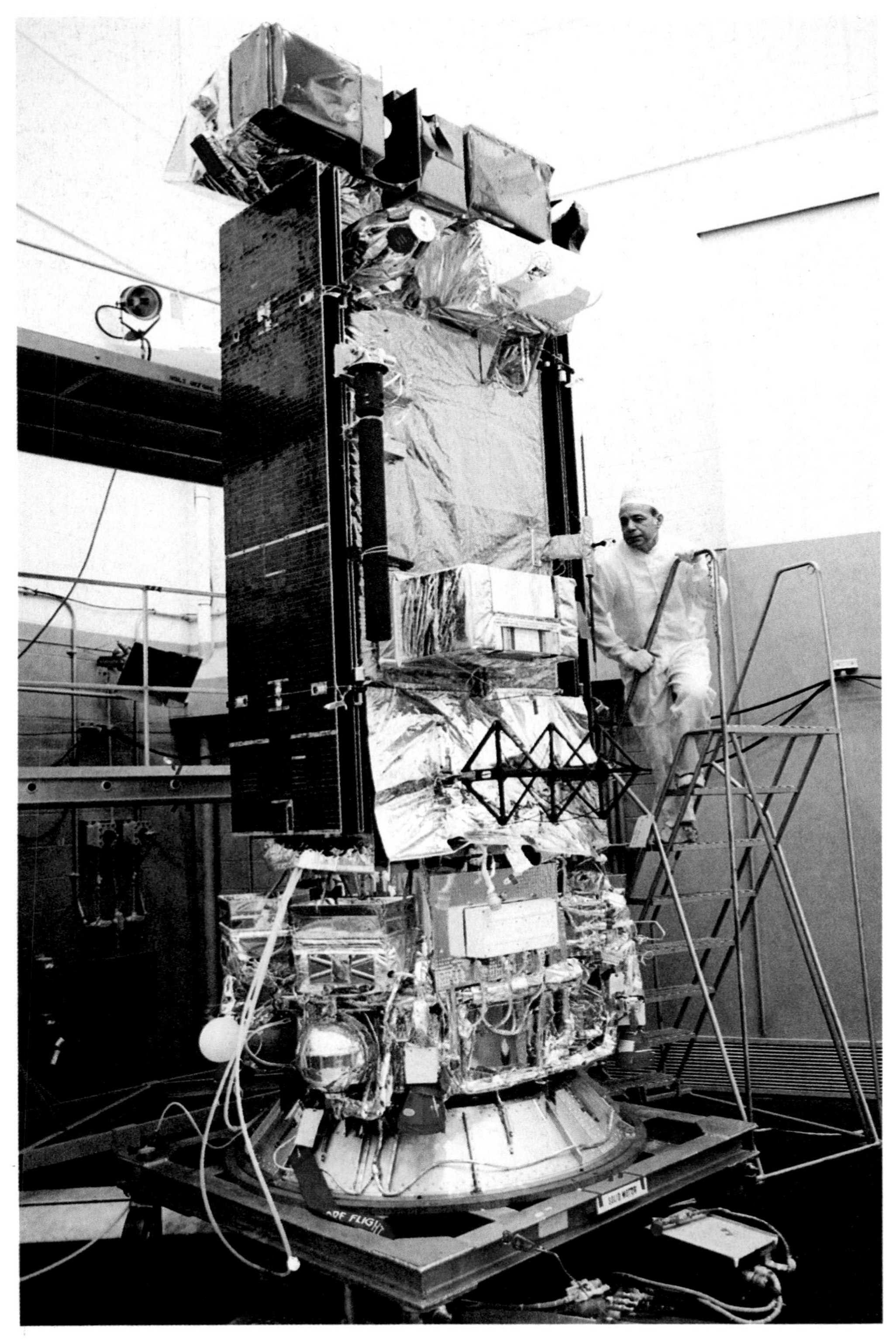

NOAA 7 had a versatile scanning radiometer.

NOAA

The NOAA-designated spacecraft were operated by NOAA. They were a development of the ITOS spacecraft, which were in turn an improvement of the Television Infrared Observation Satellite (TIROS) program. Like the ITOS spacecraft, the NOAA satellites were meteorological satellites and were developed by RCA. The first NOAA spacecraft had ITOS numbers as their primary designations (there were two NOAA 2 and NOAA 3 spacecraft because the first of each crashed). Beginning with NOAA 6, the spacecraft were larger and were based on the design of the TIROS-N-type spacecraft.
(See also ESSA, ITOS, TIROS)

NOAA 1 (ITOS-A)
(See ITOS-A)

NOAA 2 (first ITOS-B)
(See ITOS-B)

NOAA 2 (second ITOS-B)
(See ITOS-D)

NOAA 3 (first ITOS-E)
(See ITOS-E)

NOAA 3 (second ITOS-F)
(See ITOS-F)

NOAA 4 (ITOS-G)
(See ITOS-G)

NOAA 5 (ITOS-H)
(See ITOS-H)

NOAA 6
Launch vehicle: Atlas E/F
Launch: 27 June 1979
Total weight: 3097 pounds at launch
1590.6 pounds in orbit
Diameter: 74 inches
Height: 146 inches
Systems: Advanced very high resolution radiometer (AVHRR), atmospheric sounding system (ASS), and space environment monitor (SEM)

NOAA 7
Launch vehicle: Atlas F
Launch: 29 May 1980
Total weight and specifications: Same as NOAA 6
Systems: Advanced very high resolution radiometer (AVHRR), space environment monitor (SEM), solar proton monitor (SPM), TIROS operational vertical sounder (TOVS), TIROS information processor (TIP), manipulated information rate (MIR) processor and cross-strap unit (CSU)

NOAA 7C
Launch vehicle: Atlas F
Launch: 23 June 1981
Total weight and specifications: Same as NOAA 6
Systems: Same as NOAA 7

NOAA 8
Launch vehicle: Atlas F
Launch: 28 March 1983
Total weight and specifications: Same as NOAA 6
Systems: Advanced very high resolution radiometer (AVHRR) and space environment monitor (SEM), plus special search and rescue instrumentation

NOAA 9
Launch vehicle: Atlas F
Launch: 9 December 1984
Total weight and specifications: Same as NOAA 6
Systems: Same as NOAA 6

ITOS/NOAA

Spacecraft	Launch	Duration (days)	Weight (lb)	Systems
ITOS-1 (TIROS M)	23 Jan 1970	510	682	AVCS, APT, FPR, SPM
ITOS-A (NOAA 1)	11 Dec 1970	252	682	AVCS, APT, FPR, SPM
ITOS-B (NOAA 2)	21 Oct 1971	0	682	AVCS, APT, FPR, SPM
ITOS-C	–	–	–	–
ITOS-D (NOAA 2)	15 Oct 1972	837	736	FPR, SPM, VTPR, VHRR
ITOS-E (NOAA 3)	16 July 1971	0	750	FPR, SPM, VTPR, VHRR
ITOS-F (NOAA 3)	6 Nov 1973	1029	750	FPR, SPM, VTPR, VHRR
ITOS-G (NOAA 4)	15 Nov 1974	1463	750	FPR, SPM, VTPR, VHRR
ITOS-H (NOAA 5)	29 July 1976	1036	750	FPR, FPM, VTPR, VHRR
NOAA 6	27 June 1979	*	1590.6	AVHRR, ASS, SEM
NOAA 7	29 May 1980	*	1590.6	AVHRR, TOVS, SEM, SPM, TIP, MIR, CSU
NOAA 7C	23 June 1981	*	1590.6	AVHRR, TOVS, SEM, SPM, TIP, MIR, CSU
NOAA 8	28 Mar 1983	*	1590.6	AVHRR, SEM
NOAA 9	9 Dec 1984	*	1590.6	AVHRR, SEM

*Duration of still-operating spacecraft expected to match earlier durations

(See also ESSA, ITOS, TIROS)

NOVA

Launch vehicle: Scout G-1
First launch: 15 May 1981 (NOVA 1) (Navy 20)
Total weight: 368 pounds

The cylindrical main body of the NOVA had four solar arrays deployed windmill-like, a sensor extended below the main body on an accordianlike pylon, and a hydrazine motor to raise and circularize the orbit.

The NOVA program was a US Navy satellite-navigation system that was compatible with the Transit satellite navigation system, of which 15 spacecraft and three improved spacecraft had been launched at the time NOVA 1 was launched. NOVA preceded the more sophisticated NAVSTAR system. The spacecraft were developed by the Navy and launched by NASA and the US Air Force from Vandenberg AFB.

OAO

The OAO (Orbiting Astronomical Observatory) spacecraft, developed by Grumman and managed by NASA's Goddard Space Flight Center, were designed to conduct observations of interstellar space from outside the earth's atmosphere where atmospheric conditions could not interfere. The four spacecraft, with different payloads, had two outstanding successes and two failures.

OAO 1
Launch vehicle: Atlas Agena
Launch: 8 April 1966
Systems failure: 10 April 1966
Total weight: 3900 pounds
Diameter: 7 feet (plus 4-foot experiment cylinder)
Height: 10 feet (plus 10-foot experiment cylinder)
Width: 21 feet (with two solar arrays deployed)
Systems: Seven ultraviolet telescopes, a high-energy gamma ray detector (similar to Explorer 11), a low-energy gamma ray detector to survey photon sources, and a proportional counter to define and map X-ray sources

OAO 1 was an octagonal aluminum cylinder with butterfly winglike solar arrays. The primary battery overheated and shut down the spacecraft operation after two days.

OAO 2
Launch vehicle: Atlas Centaur
Launch: 7 December 1968
Last transmission: 13 February 1973
Total weight: 4446 pounds (heaviest US scientific satellite to date)
Specifications and description: Same as OAO 1
Systems: Seven ultraviolet telescopes, four large-aperture television cameras with broad-band photometers and six gimballed star trackers (each with a 3.5-inch reflecting telescope)

Launched successfully, OAO 2 survived four times longer than planned, making 22,560 observations of 1930 objects (including stars, planets and supernovae). It discovered a hydrogen cloud around the comet Tago Sato Kosaka and ultraviolet emissions from Uranus.

OAO-B
Launch vehicle: Atlas Centaur
Launch and re-entry: 30 November 1970
Total weight: 4680 pounds
Specifications and description: Same as OAO 1
Systems: Ultraviolet telescopes with a five-times greater range than those of OAO 2, and a 38-inch Cassegrain telescope

The Centaur nose shroud failed to separate and the OAO-B spacecraft fell back into the atmosphere, burning up over Africa or the Indian Ocean.

OAO 3 (Copernicus)
Launch vehicle: Atlas Centaur
Launch: 21 August 1972
Scheduled end of orbital life: 21 August 1972
Total weight: 4900 pounds
Specifications and description: Same as OAO 1
Systems: 120-inch ultraviolet telescope, 32-inch mirror, ultraviolet spectrometer and sensors for telescope guidance capable of viewing stars as faint as seventh magnitude (Princeton University Experiment Package); three smaller telescopes and a proportional counter to precisely locate X-ray sources (University College of London Experiment Package)

Within five months, OAO 3 had conducted 1780 observations of 37 objects and was declared successful. Large deuterium clouds were discovered, and the erratic behavior in the rotational period of the double star Cygnus X-1 was also exposed. The spacecraft was renamed *Copernicus* in honor of the 500th anniversary of the famous astronomer.

OFO

The OFO (Orbiting Frog Otolith) experimental program was undertaken by NASA's Office of Advanced Research and Technology, and the OFO spacecraft was developed by Aerojet General. Two bullfrogs were placed in a water-filled centrifuge with microelectrodes surgically implanted in the vestibular (inner ear) nerves leading from sensor cells in the otoliths on gravity receptors of the ear. The object of the experiment was to study the effects of weightlessness and the response to acceleration of the part of the inner ear that controls balance. Frogs were selected because their inner ear is similar to the human inner ear. It should be noted that 23 US manned space missions (including seven Apollo missions) had already been undertaken when OFO 1 was launched.

OFO 1
Launch vehicle: Scout B
Launch: 9 November 1970
Total weight: 293 pounds (including a 91-pound experiment package)
Diameter: 30 inches
Height: 47 inches

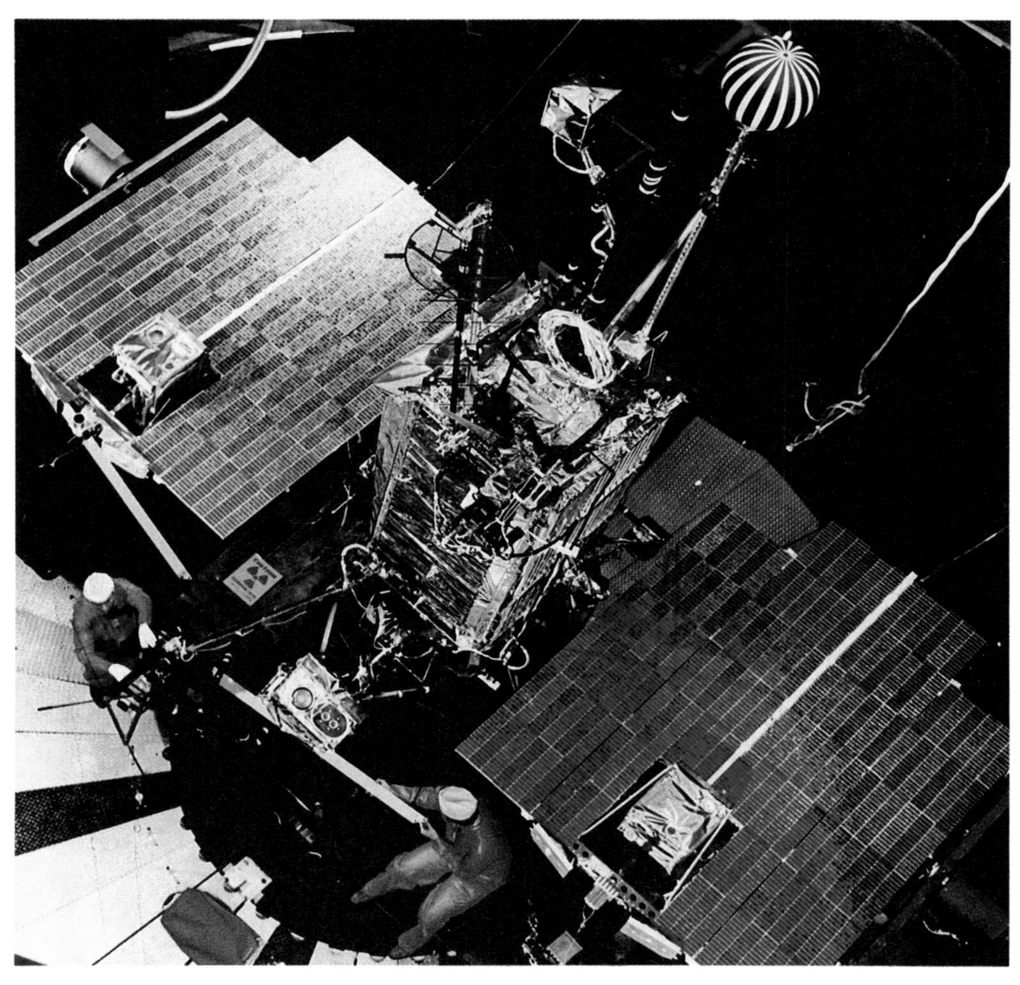

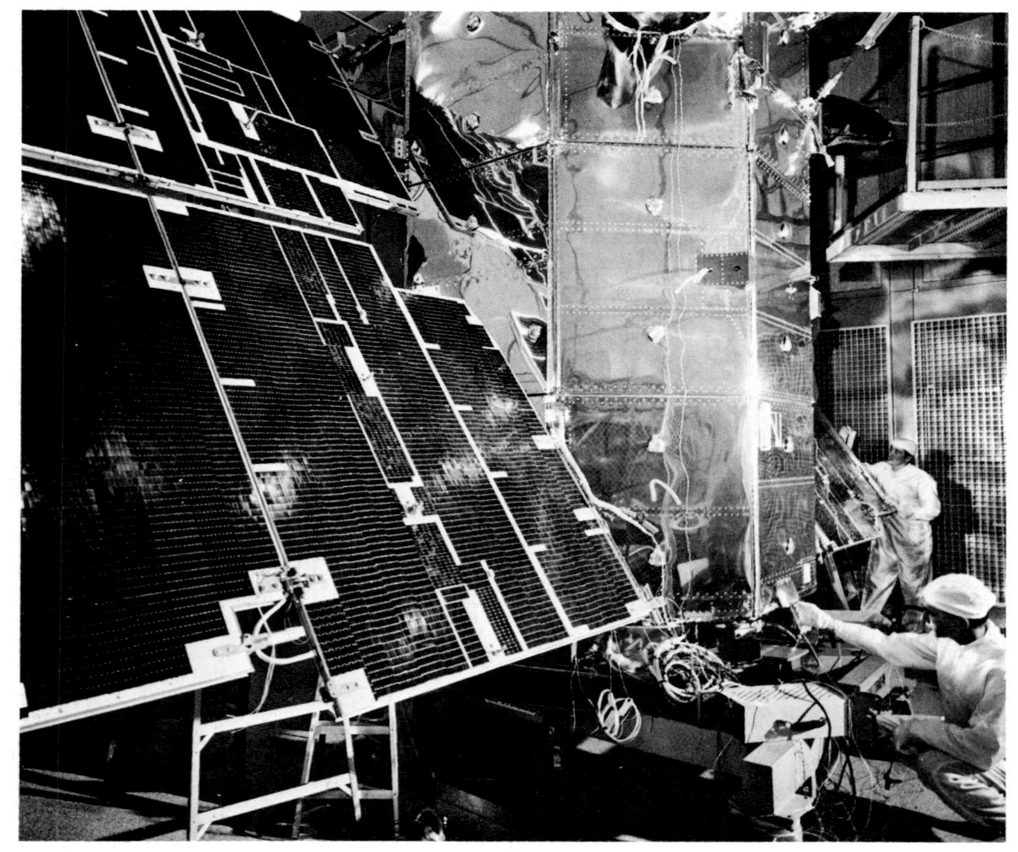

Above: **OGO 2 was fitted with 22 experiments.** ***Below:*** **OAO 2, the heaviest, largest and most complex automated spacecraft of its time, carried 11 telescopes to study the origin, evolution and future of the universe.**

The octagonal electronics section of the OFO 1 was connected by a truncated cone to an 18-inch sphere holding the OFO centrifuge, which was enclosed in a pressure-tight container. Data received over a six-day period showed that the frogs' nervous systems had difficulty adjusting to loss of gravity for the first three days, after which they functioned normally. No attempt was made to recover the frogs, and there was no OFO 2.

OGO

The OGO (Orbiting Geophysical Observatory) spacecraft were designed for NASA's Goddard Space Flight Center by TRW to conduct correlated investigations of geophysical and solar phenomena in the earth's atmosphere and magnetosphere and in interplanetary space.

OGO 1
Launch vehicle: Atlas Agena
Launch: 6 September 1964
Total weight: 1046 pounds
Diameter: 20 feet
Height: 59 feet
Shell composition: Aluminum

OGO 1 was intended to conduct studies of earth-sun interplanetary space relation-

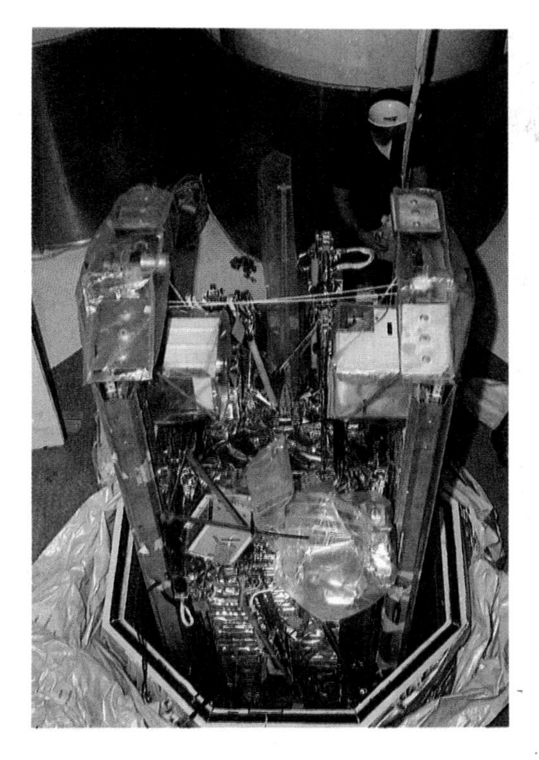

experiments), trapped radiation (two experiments), magnetic fields (two experiments), ionosphere (five experiments), optical and radio emission (three experiments), and one micrometeoroids experiment. OGO 3 was the first fully successful OGO spacecraft.

OGO 4 (OGO D, POGO)
Launch vehicle: Thor Agena D
Launch: 28 July 1967
Re-entry: 16 August 1972
Total weight: 1216 pounds
Diameter: 36 inches
Height: 72 inches
Shape: Square
Experiment booms: 22 feet (two), 4 feet (four)
Solar panels: 7 feet 6 inches by 6 feet

Far left: **The primary mission of OGO 5 was to study particles and fields in space. *Left:* OGO 6 was launched into low earth orbit. *Below:* OSO 2 studied sun structure and behavior.**

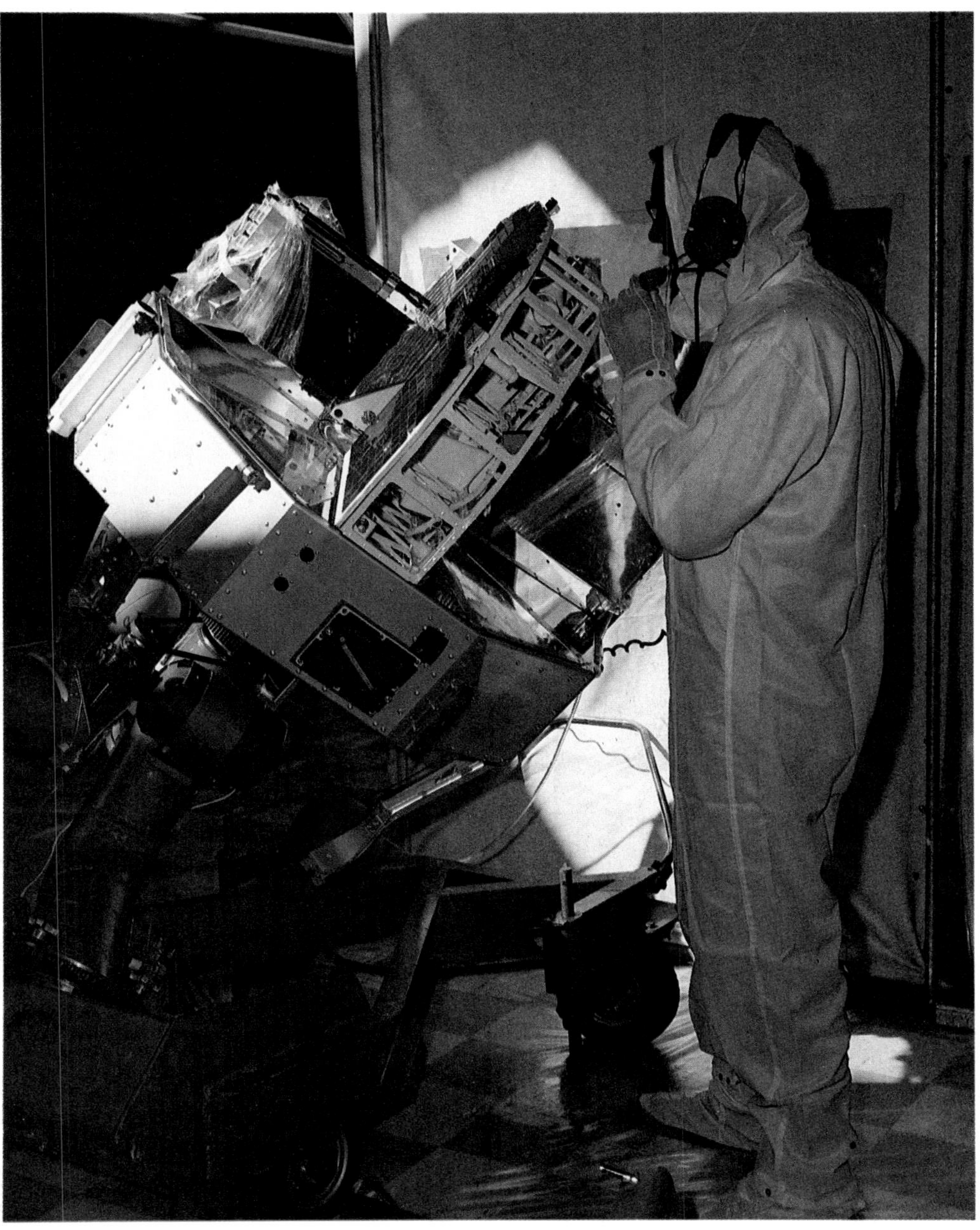

ships, energetic particles and fields, atmospheric physics and solar emissions. However, the spacecraft's booms did not deploy properly and only limited data was returned.

OGO 2
Launch vehicle: Thor Agena D
Launch: 14 October 1965
Observatory operations discontinued: 22 February 1968
Re-entry: 17 September 1981
Total weight: 1118 pounds
Diameter: 33 inches
Height: 68 inches (708 inches with booms deployed)
Width: 246 inches across solar panels
Shell composition: Aluminum
Shape: Square

This spacecraft was more successful than the similar OGO 1, concentrating on atmospheric studies and the World Magnetic Survey.

OGO 3
Launch vehicle: Atlas Agena B
Launch: 7 June 1966
Total weight: 1135 pounds
Diameter: 33 inches
Height: 68 inches
Shape: Square
Experiment booms: 22 feet (two), 4 feet (four)
Solar panels: 49 feet by 20.5 feet

OGO 3 carried scientific instrumentation totaling 195 pounds that was divided into eight areas of investigation: cosmic rays (four experiments), plasma (four

Orbital plane experiment packages: 18 inches by 102 inches

The second successful OGO carried 20 experiments, including 10 from nine universities, one from Australia and five for NASA's Jet Propulsion Laboratory.

OGO 5
Launch vehicle: Atlas Agena SLV-3A
Launch: 4 March 1968
Retired: 14 July 1972
Total weight: 1347 pounds
Specifications and description: Same as OGO 4

This spacecraft carried a record 25 experiments similar to those of earlier OGOs and including some from England, France and the Netherlands.

OGO 6
Launch vehicle: Thor AD/Agena D
Launch: 5 June 1969
Re-entry: 12 October 1979
Total weight: 1393 pounds
Specifications and description: Same as OGO 4

Launched during a period of high sunspot activity, this spacecraft studied the ionosphere and the polar auroral regions. The major emphasis was on investigating the interrelationships among particle activity, aurora and airglow, the geomagnetic field, neutral and ionized composition, wave propagation and noise, and solar energy contributing to ionization and heating.

OSCAR

The OSCAR (Orbiting Satellite-Carrying Amateur Radio) spacecraft were a series of small, relatively inexpensive, communications satellites built by amateurs from several nations, notably Australia and the US, under the auspices of the Radio Amateur Satellite Corporation (AMSAT). These satellites were, in turn, intended for use by amateur 'ham' radio operators around the world, although later spacecraft were noted for educational broadcasting. The first group of four OSCARs were launched by the US Air Force between 1961 and 1965, piggybacked with military spacecraft, while the second group of five were piggybacked aboard NASA launches between 1970 and 1981. All OSCAR spacecraft were piggybacked into space, and none was the sole payload of a launch vehicle.
(See photograph, page 82)

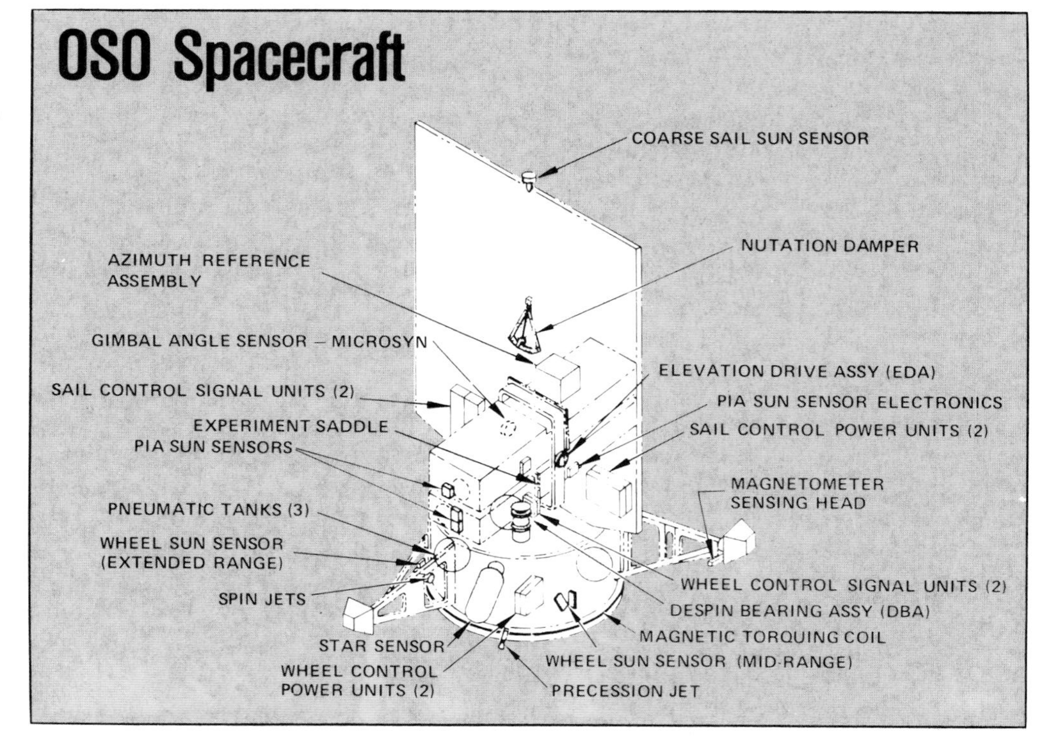

OSCAR 5
Launch vehicle: Delta N
Launch: 23 January 1970 (with ITOS 1)
Total weight: 39 pounds
Shape and dimensions: Boxlike, 12 by 17 by 6 inches

The VHF 2-meter beacon continued for 23 days, the HF 10-meter beacon lasted 46 days and the spacecraft was tracked by hundreds of amateur stations in 27 countries.

OSCAR 6
(See AMSAT-OSCAR)

OSCAR 7
Launch vehicle: Delta 2310
Launch: 15 November 1974 (with ITOS-G and the Spanish INTASAT spacecraft)
Total weight: 63 pounds

OSCAR 7 successfully provided amateur radio communication links and facilitated educational broadcasting.

OSCAR 8
Launch vehicle: Delta 2910
Launch: 5 March 1978 (with Landsat C)
Total weight: 59.4 pounds

OSCAR 8 was built by amateurs from the US, Canada, West Germany and Japan under the auspices of AMSAT and with the cooperation of the American Radio Relay League. It was launched successfully and broadcasted on the uplink frequency of 145.9 MHz (both transponders) and downlink frequencies of 29.4 MHz and 435.1 MHz.

OSCAR 9
Launch vehicle: Delta
Launch: 6 October 1981
Description: No data available

OSO

The OSO (Orbiting Solar Observatory) program spacecraft were developed for NASA's Goddard Space Flight Center by Ball Brothers Research Corp (now Ball Aerospace). The object of the program was to conduct observations of the sun and solar physics experiments investigating a broad spectral range of solar radiation. Experiments previously hindered on earth by the absorbing effects and distortions of the atmosphere were conducted from stabilized spacecraft designed to operate scientific experiments with extreme accuracy. The OSO program provided continuing study of the sun's 11-year sunspot cycle.

OSO 1 (OSO-A)
Launch vehicle: Delta
Launch: 7 March 1962
Last transmission: 6 August 1963
Re-entry: 8 October 1981
Total weight: 458 pounds
Diameter: 44 inches
Height: 37 inches
Shape: Nonagonal

OSO 1 provided data on 75 solar flares.

OSO 2 (OSO-B)
Launch vehicle: Delta DSV-3C
Launch: 3 February 1965
Last transmission: 7 October 1966

OSO 3 carried nine experiments to study the sun and its influence on the earth's atmosphere.

Total weight: 545 pounds
Diameter: 92 inches
Height: 37 inches
Shell composition: One side was faced with solar cells with a pointing/scanning experiment package mounted on that face

OSO 2 was composed of two sections: a spinning nine-sided base, or 'wheel,' section and a stabilized fan-shaped 'sail' section. Since OSO 1, this spacecraft provided data on solar flares and was able to scan the solar disk and corona.

OSO 3 (OSO-C)
Launch vehicle: Delta DSV-3C
Launch and re-entry: 25 August 1965
Total weight: 620 pounds
Diameter: 44 inches (wheel section)
Height: 15 inches

OSO 3 comprised two sections: a spinning base, or wheel, section and a top, or sail, section. The nine-sided wheel section was 44 inches in diameter and 9 inches high; 30-inch arms extended from the section with a 6-inch sphere mounted on the end of each arm. The sail section was nearly semicircular with a radius of 22 inches. Solar cells and solar pointing experiments were mounted on the side designed to face the sun.

OSO 3 failed to orbit due to launch vehicle malfunction.

OSO 3 (OSO-E)
Launch vehicle: Delta DSV-3C
Launch: 8 March 1967
Re-entry: 4 April 1982
Total weight: 627 pounds
Description: Identical to OSO-C spacecraft
Systems: Ultraviolet scanner, solar spectrometer, solar X-ray detector and telescope celestial gamma ray detector, albedo telescopes and a cosmic ray spectrum detector

This launch, like those of OSO 1 and OSO 2, was successful.

OSO 4
Launch vehicle: Delta DSV-3C
Launch: 18 October 1967
Retired: 1 November 1971
Total weight: 597 pounds
Diameter: 44 inches (101 inches with three stabilization boom arms extended)
Height: 15 inches
Systems: Ultraviolet spectrometer, X-ray spectroheliograph, X-ray spectrometer, celestial X-ray telescope, proton-electron telescope, helium and hydrogen monochrometer, geocorona H Lyman-alpha telescope and solar X-ray detector

OSO 4 consisted of a nine-sided drum that revolved at 30 rpm, allowing six experiments to scan the sun. The base of the drum held the power, command and communications subsystems. Nitrogen gas in the spheres on the arms controlled the spin rate. Solar panels and pointed instruments, mounted on a near-circular 22-inch-high sail, face the sun during the daytime. A magnetic bias coil augmented the pitch-control systems and reduced pitch-gas usage. Pointing accuracy was 0.5 arc minutes. A total of 2016 solar cells provided 38 watts of power. The maximum load was 26 watts.

All OSO 4 experiments returned useful data, including the first pictures made of the sun in extreme ultraviolet.

OSO 5
Launch vehicle: Delta
Launch: 22 January 1969
Total weight: 641 pounds
Diameter: 44 inches (92 inches with three spin-control booms extended)
Height: 38 inches

OSO 5 was constructed in the same basic two-stage design as the previous OSOs. A spinning wheel-like base and a nine-sided drum, revolving at 30 to 40 rpm, allowed five experiments housed in compartments in the base section to scan the sun every two seconds. The pitch-control system maintained the base spin axis perpendicular to the sun while the overall spacecraft was pointed within one minute of arc accuracy. The base also held the power, commands, spin-control and communication subsystems. Nitrogen gas was stored in the spheres at the end of the

booms, which were part of the controls for the spin rate. The upper section, called the sail, was fan shaped and 23 inches high, contained directionalized instrumentation and solar-cell panels, and faced the sun during the daytime. A magnetic bias coil augmented the pitch control and saved gas consumption. The design lifetime of OSO 5 was six months.

OSO 5 carried systems similar to the previous OSOs, including an extreme ultraviolet spectroheliograph and, for the first time, the ability to perform a faster scan across the solar disk. The spacecraft completed a successful mission.

OSO 6
Launch vehicle: Delta
Launch: 9 August 1969
Re-entry: 7 March 1981
Total weight: 640 pounds
Diameter: 44 inches (92 inches with spin-control booms extended)
Height: 37 inches
Description and systems: Similar to previous OSOs

OSO 6 returned successful results.

OSO 7 (OSO-H)
Launch vehicle: Delta
Launch: 29 September 1971
Re-entry: 9 July 1974
Total weight: 1400 pounds
Description: Same as previous OSOs
Systems: Similar to OSO 4

OSO 7 was successfully stabilized after a launch mishap that had sent the spacecraft tumbling into improper orbit.

OSO 8
Launch vehicle: Delta
Launch: 21 June 1975
Total weight: 2346 pounds
Diameter: 60 inches
Height: 28.2 inches
Systems: High-sensitivity crystal spectrometer and polarimeter, mapping X-ray heliometer, soft X-ray background experiment, cosmic X-ray spectroscope, high-energy celestial X-ray equipment, two ultraviolet solar telescopes, and equipment to investigate extreme ultraviolet radiation from earth and space

OSO 8 was a spin-stabilized craft with a despun platform to accommodate experiments that pointed at the sun or other sources of electromagnetic radiation. Experiments that needed no sustained pointing were carried in the rotation portion. OSO 8 had a designed lifetime of 12 months. The bottom section, called the wheel, was a cylinder 60 inches in diameter by 28.2 inches in height. The rotating wheel accounted for 70 percent of the total spacecraft weight, including 555 pounds for the six experiments located there. The nonspinning sail solar cells provided power and a platform that always faced the sun for the two pointing experiments. The sail was 92.4 inches high and 82.5 inches wide. The two pointed experiments composed a package about 15 inches square and 57 inches long that weighed 268 pounds. Solar cells and nickel-cadmium batteries provided 110 watts of power to the spacecraft and experiments. The wheel spin rate was 6 rpm and was controlled by a pneumatic system.

OSO 8, the last OSO, turned its attention to X-ray sources in the Milky Way and beyond while it continued solar physics studies.

A Hughes Aircraft Company engineer checks OSO 8's two spectrometers during testing. These instruments make fine structure studies of the chromosphere and take high-resolution ultraviolet spectrometer measurements.

PAGEOS

Launch vehicle: Thor Agena
Launch: 24 June 1966
Disintegrated: 20 January 1976 (reasons unknown)
Total weight: 125 pounds
Diameter: 100 feet
Shell composition: 84 panels of 0.0005-inch mylar coated on the outside with vapor-deposited aluminum
Shape: Spherical

The PAGEOS (Passive GEOS) spacecraft was fabricated by the GT Schjeldahl Company, with its canister and adapter systems developed by Goodyear Aerospace. It was designed as a passive Echo-type sunlight reflector. It was to be photographed over a five-year period to determine the precise location of continents, islands and other land masses and to determine geographic points in relation to one another. PAGEOS was launched from Vandenberg AFB and photographed by the stations of a worldwide triangulation network. As a passive spacecraft, PAGEOS carried no instruments, but as a reflector it could be used to pinpoint a location on earth as measured from the center of mass with a margin of error of 51 to 96 feet.
(See also Echo, Explorers 22 to 36, GEOS, LAGEOS)

Pegasus

Launch vehicle: Saturn 1
Launch and re-entry: See table below
Systems: 208 detector panels (20 by 40 inches) mounted on wings, composed of polyurethane foam core with aluminum/mylar/copper laminated skin 0.0015-inch thick (16 panels), 0.008-inch thick (32 panels) and 0.016-inch thick

The Pegasus spacecraft were developed for NASA by Fairchild Hiller to investigate micrometeoroids in advance of the Apollo manned lunar program. They were launched along with boiler-plate Apollo capsules. Because of the number of micrometeoroids discovered by Pegasus (fewer than expected), a total of 1000 pounds was able to be saved in the weight of the operational Apollo spacecraft.

In orbital configuration Pegasus consisted of two large 'wings,' each 14 by 48 feet. Four solar panels extended from the open-truss rectangular aluminum center structure mounted on top of the second stage.

The Pegasus Program

	Pegasus 1	Pegasus 2	Pegasus 3
Launch	16 Feb 1965	25 May 1965	30 July 1965
Data collection terminated	13 Jan 1968	14 Mar 1968	29 Aug 1968
Re-entry	17 Sept 1978	3 Nov 1979	4 Aug 1969
Total weight	23,000 lb	23,100 lb	23,100 lb
Scientific experiment weight	3126 lb	3200 lb	3200 lb

Pioneer

The Pioneer program, like the Explorer program, involved several dissimilar series of spacecraft for interplanetary scientific exploration. The first series of Pioneers were designed for lunar flyby and lunar orbital experiments, the second for solar orbital experiments, and the last group of Pioneers were planetary probes directed at Jupiter, Mars and Venus.

Pioneer 1
Launch vehicle: Thor Able
Launch: 11 October 1958
Re-entry: 12 October 1958
Total weight: 75.3 pounds
Diameter: 29 inches
Height: 30 inches
Shell composition: Laminated plastic
Shape: Double toroidal

The first NASA spacecraft, Pioneer 1, failed to reach the moon, but it returned 43 hours of data before crashing into the South Pacific.

Pioneer 2
Launch vehicle: Thor Able
Launch and re-entry: 8 November 1958
Total weight: 75.3 pounds
Diameter: 29 inches
Height: 30 inches
Shell composition: Laminated plastic
Shape: Double toroidal

Pioneer 2 failed to achieve orbit, but returned data on the equatorial radiation field and micrometeoroids.

Pioneer 3
Launch vehicle: Juno 2
Launch: 6 December 1958
Re-entry: 7 December 1958
Total weight: 12.95 pounds
Diameter: 10 inches
Height: 23 inches
Shell composition: Gold-washed fiber-glass painted with white stripes to control internal temperature
Shape: Conical

Pioneer 3 did not reach the moon as planned, but did discover a second radiation belt around the earth.

Pioneer 4
Launch vehicle: Juno 2
Launch: 3 March 1959
Lunar encounter: 4 March 1959
Total weight: 13.4 pounds
Diameter: 9 inches
Height: 20 inches
Shape: Conical

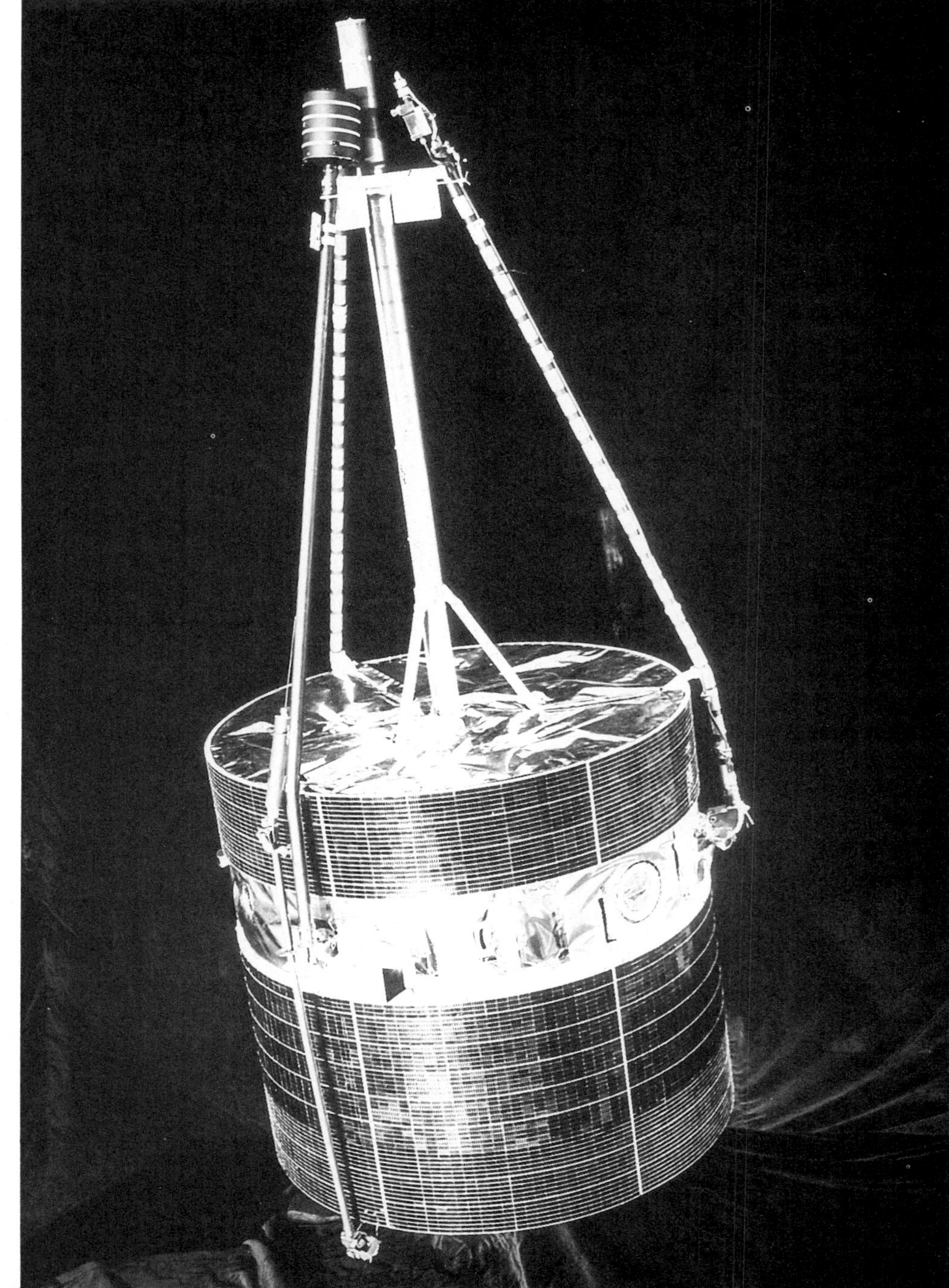

Above: **Pioneer 6 with booms in launch position.**
Right: **Pioneer 5 was launched from Cape Canaveral.**

Pioneer 4 passed within 37,300 miles of the moon at 4490 mph, but not close enough (by 20,000 miles) to trigger the photoelectric lunar radiation sensor.

Pioneer P-3
Launch vehicle: Atlas Able
Launch and re-entry: 26 November 1959

The Pioneer P-3 intended lunar orbiter was lost shortly after launch due to launch vehicle malfunction.

Pioneer 5 (P-2)
Launch vehicle: Thor Able

Right: **Test model of Pioneer 10 on the shake table to test its ability to withstand vibrations.**

Launch: 11 March 1960
Placed in solar orbit: 10 August 1961
Total weight: 94.8 pounds
Diameter: 26 inches
Shell compositon: Aluminum alloy
Shape: Spherical

Pioneer P-2 returned important data on solar flares, solar wind and galactic cosmic rays. It also established a record of 22.5 million miles for radio communication in deep space.

Pioneer P-30
Launch vehicle: Atlas Able-5A
Launch and re-entry: 25 September 1960
Total weight: 387 pounds
Diameter: 39 inches
Shell composition: Aluminum alloy
Shape: Spherical

Pioneer P-30, an intended lunar orbiter, was lost shortly after launch due to launch vehicle malfunction.

Pioneer P-31
Launch vehicle: Atlas Able-5A
Launch and re-entry: 15 December 1960
Total weight and specifications: Same as Pioneer P-30
Results: Same as Pioneer P-30

Pioneer 6
Launch vehicle: Thor Delta (DSV-3E)
Launch: 16 December 1965
Total weight: 140 pounds
Diameter: 37 inches
Height: 35 inches
Shell composition: Aluminum
Shape: Cylindrical

Placed into solar orbit after 44 hours, Pioneer 6 returned data on solar wind and magnetic fields.

Pioneer 7
Launch vehicle: Delta
Launch: 17 August 1966
Total weight: 140 pounds
Diameter: 37 inches
Height: 37 inches
Shell composition: Aluminum

A 52-inch mast containing high-gain and low-gain antennae projected from one end of the spacecraft and dual-frequency antennae used in radio propagation experiments were deployed from the other end. Three 64-inch booms extended from the midsection, with a magnetometer, wobble damper and orientation nozzle on the ends of their respective booms. The sides of the cylinder were covered with solar cells.

Placed in elliptical solar orbit, Pioneer 7 continued the program of measuring solar wind at widely separated points.

Pioneer 8
Launch vehicle: Delta DSV-3E
Launch: 13 December 1967
Total weight: 44 pounds
Diameter: 37 inches
Height: 35 inches
Shell composition: Aluminum
Shape: Cylindrical
Results: Same as Pioneer 7

Pioneer 9
Launch vehicle: Delta (DSV-3E)
Launch: 8 November 1968
Total weight: 148 pounds
Diameter: 37 inches
Height: 35 inches
Shell composition: Aluminum

Three 64-inch booms were deployed at 120-degree intervals from the center of one end, and experiment antennae were mounted on the rim of the other end. Spin was stabilized at 60 rpm. The sides of the spacecraft were covered with solar cells, except for a narrow circular band around the middle containing apertures for four experiments and five sun sensors.

Pioneer 9 achieved the same as Pioneers 6 through 8.

Pioneer E
Launch vehicle: Delta
Launch and re-entry: 27 August 1969
Total weight: 45 pounds
Diameter: 37 inches
Height: 35 inches
Shell composition: Aluminum
Shape: Cylindrical

Intended for a solar mission similar to those of Pioneers 6 through 9, this spacecraft was destroyed shortly after launch due to launch vehicle malfunction.

Pioneer 10 (Pioneer F)
Launch vehicle: Atlas Centaur
Launch: 3 March 1972
Asteroid belt encounter: 15 July 1972
Jupiter encounter: 3 December 1973
Total weight: 570 pounds
Diameter: 9 feet (equipment compartment)
Height: 14 inches (equipment compartment)
Systems: Spacecraft radio, meteoroid detector, meteoroid/asteroid detector, plasma analyzer, helium vector magnetometer, charged particle detector, cosmic ray telescope, geiger-tube telescope, trapped-radiation telescope, ultraviolet photometer, infrared radiometer and an imaging photopolarimeter

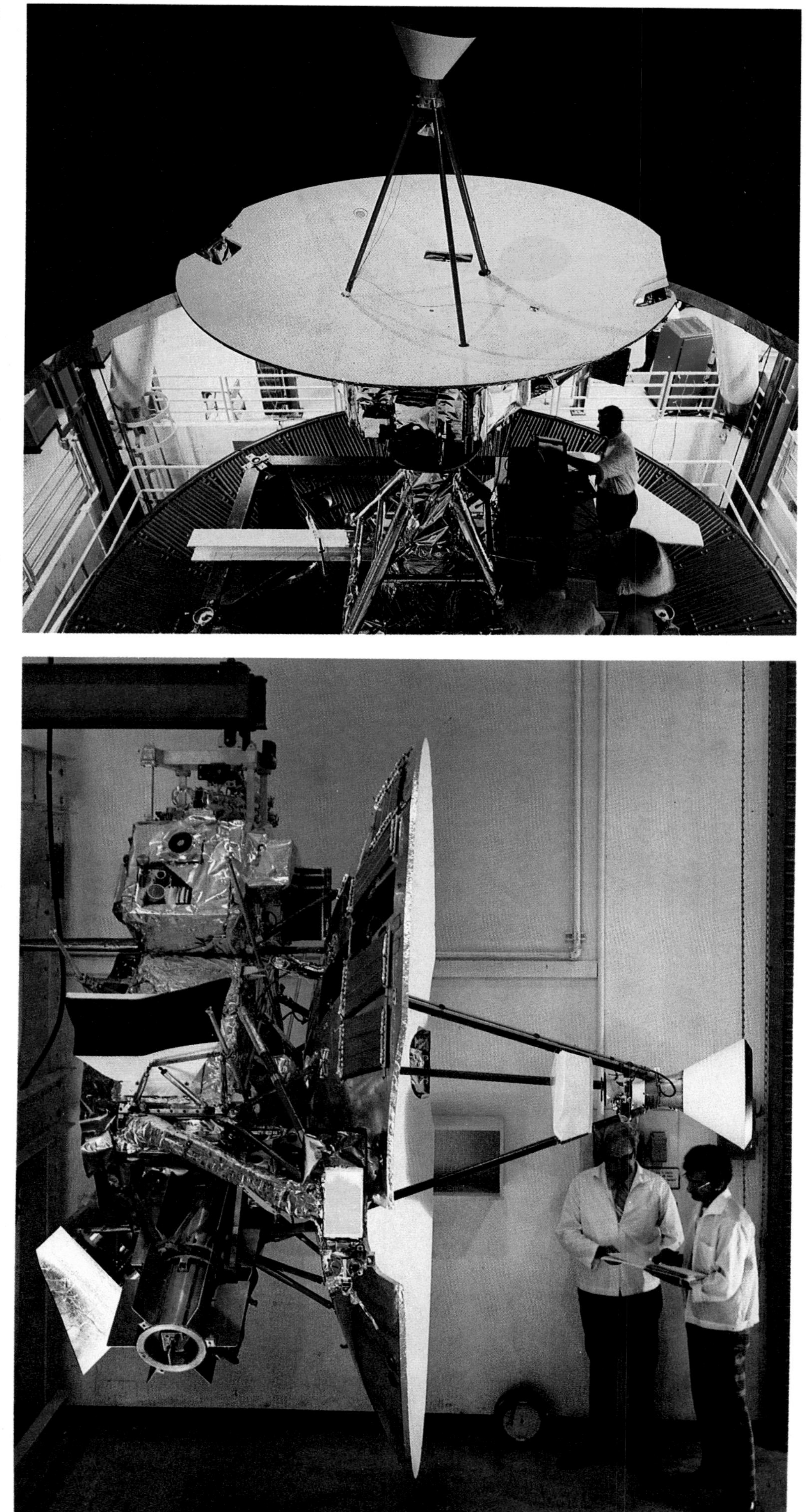

Pioneer 10, like Pioneer 11, was designed to conduct the first flyby of one of the outer planets, Jupiter, and then continue indefinitely into deep space. Despite fears that it would strike an asteroid and be put out of commission, the spacecraft successfully negotiated the asteroid belt between 15 July 1972 and 15 February 1973. During August 1973 it participated with Pioneer 9 in observing one of the most severe solar storms ever recorded. On 8 November 1973 the spacecraft passed the orbital path of Jupiter's outermost moon, Hades. On 26 November, the spacecraft's cameras were turned on and more than 300 photographs were returned of Jupiter and its moons, Callisto, Europa and Ganymede. The closest approach to Jupiter itself (81,000 miles) took place on 3 December. Though its trajectory crossed the orbital paths of Saturn, Uranus, Neptune and Pluto, Pioneer 10 did not pass close enough to these planets to return any data. The spacecraft is now out of communications range of earth, and in 1987 it will be the first man-made object to leave the solar system. Pioneer 10 is scheduled to reach the vicinity of the star Aldebaran in approximately the year AD 8,001,972.

Like its successor, Pioneer 11, Pioneer 10 carried a plaque into the universe containing symbols that are expected to be decipherable by intelligent life, should it happen to encounter a Pioneer (see illustration). The symbols include (counterclockwise from upper left) a symbol for the hydrogen atom, one of the most common in the universe; a symbolic map of the radio energy originating from our sun so that the stellar source of Pioneer can be identified; a map of the solar system

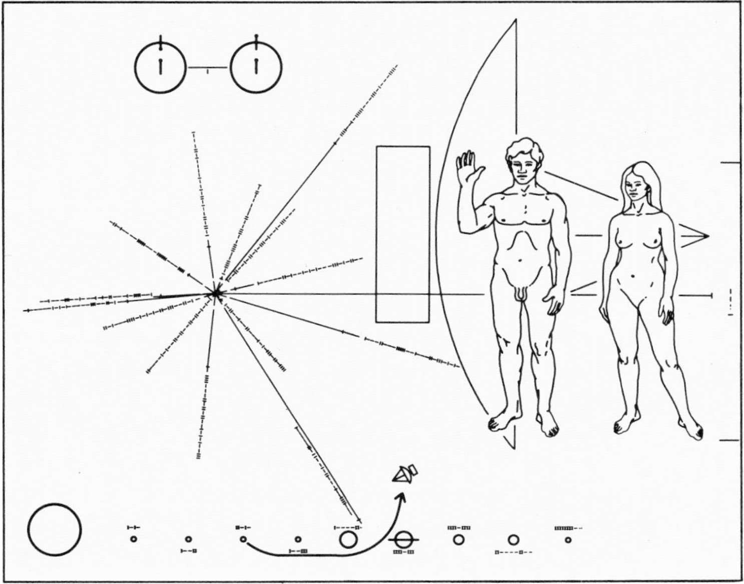

Left: **Pioneer 10 is raised into the space simulation chamber** ***above*** **and mounted on the test stand for calibration and checkout,** ***below. Right:*** **Pioneer 11 photographed Jupiter's satellite Io from a distance of more than a million miles (*top*).** ***Above:*** **Pioneers 10 and 11 carried this plaque into interstellar space.**

showing the location of earth with Pioneer leaving it; and stylized drawings of male and female human beings superimposed over a stylized drawing of Pioneer in the same scale.
(See also Pioneer 11, Voyager)

Pioneer 11 (Pioneer G)
Launch vehicle: Centaur
Launch: 6 April 1973
Jupiter encounter: 3 December 1974
Saturn encounter: 1 September 1979
Total weight and specifications: Same as Pioneer 10

Pioneer 11 carried the same systems as Pioneer 10, plus a flux-gate magnetometer. Its mission was very similar to that of Pioneer 10, although it was designed for a closer approach to Jupiter and a flyby of Saturn. Like its predecessor, it safely passed through the asteroid belt and continued toward the outer planets, passing 26,680 miles below Jupiter's south pole on 3 December 1974, exactly a year after Pioneer 10's closest approach. On 20 November 1975 it made its first observations of Saturn and, although it was still far from the ringed planet, these were the nearest observations yet made by a man-made observation platform. Even though Pioneer 11 was traveling at roughly 39,700 mph, faster than any man-made object yet, it took four years and nine months to cover the distance from Jupiter to Saturn. In September 1979, more than six years after launch, Pioneer 11 passed within 21,000 miles of Saturn's rings, returning considerable valuable data about the ring system. After passing Saturn, Pioneer 11 plunged into deep space, carrying with it a plaque similar to that aboard Pioneer 10 in the hope that someday another form of intelligent life will find it.

Although they were to be upstaged by the two Voyager spacecraft already on their way, Pioneers 10 and 11 were the true pioneers of the outer solar system, providing data about the outer planets that was many times greater and more detailed than had been collected in all the years before them.
(See also Pioneer 10, Voyager)

Pioneer Venus 1 (Orbiter)
Launch vehicle: Atlas Centaur
Launch: 20 May 1978
Injected into orbit around Venus: 4 December 1978
Total weight: 1283 pounds (including 99 pounds of scientific instruments) at launch; 811 pounds in orbit
Diameter: 8 feet

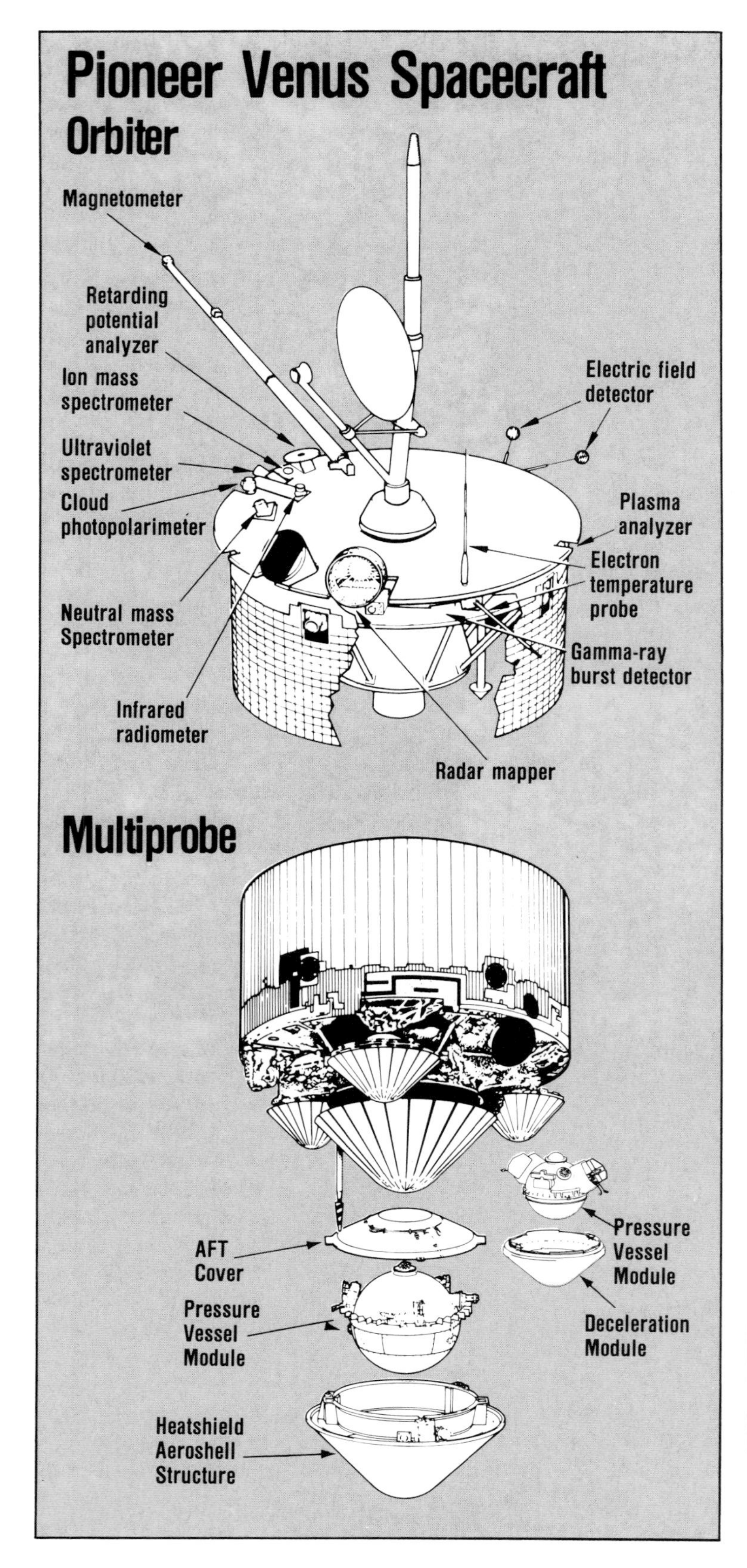

Top: **Pioneer Venus orbiter and multiprobe spacecraft (rear).** ***Above:*** **Pioneer Venus Project manager C Hall with the multiprobe thermal model.**

Height: 4 feet
Systems: Cloud photopolarimeter, surface radar mapper, infrared radiometer, airglow ultraviolet spectrometer, ion-mass spectrometer, solar wind plasma analyzer, magnetometer, electric-field detector, electron temperature probes, charged-particle retarding potential analyzer, and gamma ray burst detector

Pioneer Venus-A (orbiter) and Pioneer Venus-B (multiprobe) spacecraft shared a basic modular design. The main bodies of both were flat cylinders. These provided a spin-stabilized platform for scientific instruments and spacecraft systems. A solar array was attached to the equipment shelf. The Venus orbiter spacecraft incorporated a despun, high-gain dish antenna on a 10-foot mast and sensor elements mounted on booms. Magnetometer sensors were mounted on a three-section, deployable 15-foot boom.

The Pioneer Venus orbiter completed its 300-million-mile voyage in just over six months and was placed into Venusian orbit on 4 December 1978, less than a week before arrival of the Pioneer Venus multiprobe. Its orbital path took it as close as 93 miles to the planet's surface, sampling its upper atmosphere. The orbiter's primary mission was to observe the planet for the length of time it takes for Venus to make one revolution on its axis (243 earth days, or 1 Venusian day). Observations included operation of all the spacecraft's systems and daily infrared and ultraviolet pictures of the planet's cloud cover.

Pioneer Venus 2 (Multiprobe)
Launch vehicle: Atlas Centaur
Launch: 8 August 1978
Venus landing: 9 December 1978
Total weight: 1993 pounds (including 112 pounds of scientific instruments)
Diameter: 8 feet (at widest point)
Systems: Neutral mass spectrometer; gas chromatograph; solar flux radiometer; infrared radiometer; cloud-particle size spectrometer; nephelometer and pressure/temperature/acceleration sensors (large probe); nephelometer; net flux radiometer; and pressure/temperature/acceleration sensors (each of three small probes)

Pioneer Venus 2, a multiprobe, was made up of a transporter bus, a large probe and three identical smaller probes. All were to enter the atmosphere of Venus at various places, and all carried instruments and sensors to measure atmospheric and surface phenomena. The transponder was a spin-stabilized, short cylinder 8 feet in diameter that housed instruments, communications and navigation systems. All four probes were geometrically similar. The main component of each housed instruments, communications, data, command and power systems. The large probe weighed 697 pounds and was 5 feet in diameter; it seven instruments weighed 62 pounds. The smaller north, day and night probes (named for their destination on the planet's surface) each weighed 205 pounds. The four probes were launched from the multiprobe 8 million miles from Venus to fly to their entry points, two on the day side and two on the night side of Venus. The probe pressure vessels were made of titanium for pressure and heat resistance. Windows for infrared radiometer instruments consisted of diamonds (13.5 carats on the large probe) cut from a 205-carat industrial-grade rough diamond.

The Pioneer Venus multiprobe lander conducted the only landing on the surface of Venus by a US spacecraft, although it was preceded by two and followed by four successful Venusian landings by Venera spacecraft from the USSR. It was the only Pioneer designed to land on another planet. The Pioneer Venus 2 probes separated as they entered the Venusian atmosphere to descend far apart. The probes were designed as hard landers, that is, they were intended to return data only up to their point of impact on the surface of Venus and not equipped with systems that would break their fall and permit a soft landing. Nevertheless, one of the landers survived the crash into the surface of the planet and returned data for 67 minutes.

Probe A/Probe B

Launch vehicle: Scout
Launch: 19 October 1961 (Probe A)
29 March 1962 (Probe B)
Re-entry: 19 October 1961 (Probe A)
29 March 1962 (Probe B)
Suborbital distances: 4261 miles (Probe A)
3910 miles (Probe B)

Launched from Wallops Island, the two suborbital scientific geoprobes were designed to measure electron density and to test the Scout launch vehicle.

Project 78

Launch vehicle: Atlas
Launch: 24 February 1979
Total weight: 1944 pounds

The Project 78 spacecraft was developed by Ball Aerospace for the US Air Force. It is a dual-spin earth/sun observation platform that measures gamma rays and solar dynamics.

Radiation Meteoroid Satellite

Launch vehicle: Scout B
Launch: 9 November 1970 (with OFO)
Re-entry: 7 February 1971
Total weight: 46.3 pounds
Diameter: 30 inches
Height: 15 inches (solar cylinder)
Shape: Cylindrical

The Radiation Meteoroid Satellite (RMS) was developed by Ling-Temco-Vought to measure impact flux and the direction and speed of meteoroids in order to develop safety systems for future manned spaceflights. It carried an advanced dosimetry system with a solid-state radiation spectrometer.

Radio Astronomy Explorer

(See Explorers 38, 49)

RAM Re-entry Experimenters

Launch vehicle: Scout
Launch: 19 October 1967 (C-1)
22 August 1968 (C-2)
30 September 1970 (C-3)
Re-entry: 19 October 1967 (C-1)
22 August 1968 (C-2)
30 September 1970 (C-3)
Total weight: 255.2 pounds (C-1) to 294.8 pounds (C-3)
Re-entry velocity: 25,000 fps

The RAM program tested the effects of re-entry velocity on communications, and in the C-3 mission compared the effectiveness of freon to water in alleviating radio blackout during re-entry.
(See also Re-entry)

Ranger

The Ranger program was the first of three intermediate steps leading up to the Apollo program that placed a human being on the moon. The Ranger spacecraft were designed to fly to the moon, returning data enroute and up to their point of impact on the surface. The point of impact was designed to consist of a hard landing, or crash, with no provision for preserving the spacecraft through the crash. The spacecraft were designed to transmit television pictures of their approach, but it was not until Ranger 7 that the program was successful. The Ranger program was followed by the lunar orbiters and later by the Surveyor program that included soft landings on the moon in which the spacecraft were designed to survive. The Ranger spacecraft were designed by NASA's Jet Propulsion Laboratory in conjunction with the California Institute of Technology.
(See also Apollo, Lunar Orbiter, Surveyor)

Ranger 1
Launch vehicle: Atlas Agena
Launch: 23 February 1961
Re-entry: 30 August 1961
Total weight: 675 pounds
Diameter: 5 feet
Height: 13 feet
Shell composition: Gold plating, white paint and polished aluminum
Shape: Conical

Ranger 1's earth orbital, prelunar mission was only marginally successful, with low orbit due to Agena malfunction.

Left and right: **Ranger 1 was launched into low parking orbit but did not attain its eccentric orbit.**

Ranger 2
Launch vehicle: Atlas Agena
Launch and re-entry: 18 November 1961
Total weight: 675 pounds
Diameter: 17 feet (across solar panels)
Height: 13 feet
Shell composition: Gold plating, white paint and aluminum surfaces
Shape: Conical
Results: Same as Ranger 1

Ranger 3
Launch vehicle: Atlas Agena
Launch: 26 January 1962
Total weight: 727 pounds
Diameter: 23 inches
Width: 17 feet (across solar panels)
Height: 123 inches
Shell composition: Silvered plastic sheet

Ranger 3 failed to land on the lunar surface, missing by 22,862 miles on 28 January 1962, but it provided engineering data for future Rangers.

Ranger 4
Launch vehicle: Atlas Agena
Launch: 23 April 1962
Impact on the moon: 26 April 1962
Total weight: 730 pounds
Specifications and description: Same as Ranger 3

Equipment failure rendered the control and other systems of Ranger 4 useless, but the inert spacecraft reached the moon, impacting on the far side.

Ranger 5
Launch vehicle: Atlas Agena
Launch: 18 October 1962
Impact on the moon: None
Total weight: 755 pounds
Specifications and description: Same as Ranger 3

The solar cells on Ranger 5 failed to recharge the batteries. The spacecraft missed the moon by 450 miles and no pictures were returned.

Ranger 6
Launch vehicle: Atlas Agena
Launch: 30 January 1964
Impact on the moon: 2 February 1964
Total weight: 840 pounds
Diameter: 15 feet (across solar panels)
Height: 123 inches
Shell composition: Aluminum
Shape: Hexagonal

Ranger 6 achieved lunar impact within 20 miles of its target, but the camera system failed and no pictures were returned.

Ranger 7
Launch vehicle: Atlas Agena
Launch: 28 July 1964
Impact on the moon: 31 July 1964
Total weight: 806 pounds
Specifications and description: Same as Ranger 6

Ranger 7 was the first successful Ranger. It returned 4316 television pictures with resolution down to 3 feet, and it landed in the Sea of Clouds, less than 10 miles from its intended landing site.

Ranger 8
Launch vehicle: Atlas Agena
Launch: 17 February 1965
Impact on the moon: 20 February 1965
Total weight: 809 pounds (including a 380-pound camera structure)
Specifications and description: Same as Ranger 6

Ranger 8 returned 7137 television pictures with resolution down to 5 feet. It landed in the Sea of Tranquility, about 15 miles from its intended landing site.

Ranger 9
Launch vehicle: Atlas Agena
Launch: 21 March 1965
Impact on the moon: 24 March 1965
Total weight: 809 pounds (including a 380-pound camera structure)
Specifications and description: Same as Ranger 6

Ranger 9, the last Ranger, returned 5814 television pictures with resolution down to 10 inches, 200 of which were broadcast live on television. The spacecraft landed in the crater Alphonsus, 2.76 miles from its intended landing site, which was also in the crater.

10 9
11
NO STEP

RCA Satcom

The Radio Corporation of America (RCA) began orbiting commercial communications and later cable-television satellites under the RCA designation in 1975. The RCA spacecraft were built by RCA's Astro Eletronics Division, owned by RCA American and managed by RCA Global Communications Incorporated. The first three spacecraft were designed as a unit to provide communications coverage of Alaska, Hawaii and the continental United States from geostationary orbit.

RCA Satcom A (RCA Satcom 1)
Launch vehicle: Delta 3914
Launch: 13 December 1975
Total weight: 1913 pounds at launch
1021 pounds in orbit
Transponders: 24 for television, FM radio, digitized data and voice (36 MHz)
Voice capability: 24,000 circuits
Digitized data capability: 64 million bits per second
Frequency: 4-6 GHz
Position: 135° West longitude

The first RCA Satcom was box-shaped and measured 5.3 feet by 4.1 feet by 4.1 feet. Two rectangular solar panels extended from the spacecraft on short booms. The 75.5 square feet of silicon solar cells were continuously oriented toward the sun to provide 770 watts of power. The spacecraft was three-axis stabilized and earth oriented.

RCA Satcom B (RCA Satcom 2)
Launch vehicle: Delta 3914
Launch: 26 March 1976
Total weight: 1900 pounds at launch
1020 pounds in orbit
Systems and capability: Same as RCA Satcom A
Position: 128° West longitude

The main structure was box shaped, 5.3 by 4.1 by 4.1 feet, with its antenna mounted on top. Two winglike solar arrays extended from the sides of the structure to provide about 770 watts of power. An apogee kick motor was mounted in the base of the main body.

RCA Satcom C (RCA Satcom 3)
Launch vehicle: Delta 3914
Launch: 6 December 1979
Total weight: 1800 pounds at launch
1000 pounds in orbit

Ranger 2's mission was to test the spacecraft, not to land on the moon.

Systems and capability: Same as RCA Satcom A

The box-shaped main structure had dimensions of 47 by 64 by 46 inches. Two 61-inch-by-89-inch soar paddles extended in winglike fashion and provided about 770 watts of electrical power. A four-reflector antenna assembly was mounted on top of the main body.

Communications with the RCA Satcom C were lost shortly after launch.

RCA Satcom D (RCA Satcom 3R)
(Replacement for RCA Satcom 3)
Launch vehicle: Delta 3910
Launch: 19 November 1981
Total weight: 2385 pounds
Systems and capabilities: Same as RCA Satcom A
Position: 131° West longitude

With solar panels deployed, the satellite spanned 37 feet. The spacecraft main body measured 64 by 50 by 51 inches.

RCA Satcom 4
Launch vehicle: Delta
Launch: 15 January 1982
Description: Similar to earlier Satcoms
Systems and capabilities: Same as earlier Satcoms
Position: 83° West longitude

RCA Satcom 5
Launch vehicle: Delta
Launch: 27 October 1982
Description: Similar to earlier Satcoms
Systems and capabilities: Same as earlier Satcoms

RCA Satcom F (RCA Satcom 1R)
(Replacement for RCA Satcom 1)
Launch vehicle: Delta
Launch: 11 April 1983
Description: Similar to earlier Satcoms
Systems and capabilities: Same as earlier Satcoms
Position: 139° West longitude

RCA Satcom G (RCA Satcom 2R)
(Replacement for RCA Satcom 2)
Launch vehicle: Delta
Launch: 8 September 1983
Description: Similar to earlier Satcoms
Systems and capabilities: Same as earlier Satcoms
Position: 66° West longitude

Re-entry

Launch vehicles: Scout
Launch: 1 March 1962 (R-1)
31 August 1962 (R-2)

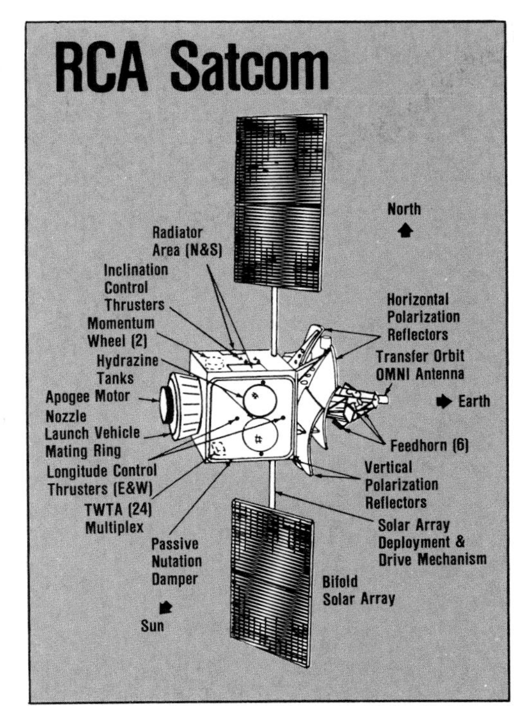

20 July 1963 (R-3)
18 August 1964 (R-4)
9 February 1966 (R-5)
27 April 1968 (R-6)
Re-entry: 1 March 1962 (R-1)
31 August 1962 (R-2)
20 July 1963 (R-3)
18 August 1964 (R-4)
9 February 1966 (R-5)
27 April 1968 (R-6)
Total weight: Ranged from 209 pounds (R-5) to 598.4 pounds (R-6)

The re-entry test program spacecraft were launched from Wallops Island in a series of suborbital tests to evaluate various types of heat ablative materials and advance the development of atmospheric re-entry technology, particularly with regard to the impending Apollo program. The desired 28,000 fps speed was not achieved on either R-1 or R-2, and R-6 performance was nominal. The R-4 spacecraft demonstrated a low-density charring ablator at 17,950 fps re-entry and the R-5 (Re-entry E) spacecraft evaluated the char integrity of a low-density phenolic-nylon ablator at 27,000 fps. (See also RAM Re-entry Experimenters)

Relay

Launch vehicle: Thor Delta
Launch: 13 December 1962 (Relay 1)
21 January 1964 (Relay 2)
Last transmission: 2 February 1965 (Relay 1)
23 May 1964 (Relay 2)
Total weight: 172 pounds
Diameter: 29 inches (at base)

The eight-sided Relay communication satellite was studded with 8215 solar cells.

Height: 32 inches (with 18-inch wide-band antenna)
Receiving frequency: 1723.3-1727 Mc
Transmission frequency: 4165-4175 Mc
Channels: 300 one-way voice or 12 two-way narrow band

The Relay spacecraft were designed by RCA and launched by NASA to test intercontinental microwave communications with a low-altitude repeater satellite and to measure radiation levels. The spacecraft, capable of both television and voice communications, were linked with stations in the United States, Europe, Japan (Relay 2 only) and Brazil.
(See also RCA Satcom)

RFD

Launch vehicles: Scout
Launch: 22 May 1963 (RFD 1)
9 October 1964 (RFD 2)
Re-entry: 22 May 1963 (RFD 1)
9 October 1964 (RFD 2)
Total weight: 478.72 pounds

The RFD, or Re-entry Flight Demonstrator, tests involved placing US Atomic Energy Commission (AEC) nuclear reactor mock-ups in space on a suborbital trajectory and evaluating re-entry effects.

The tests were conducted at a time when NASA and AEC were working toward the development of a nuclear rocket engine under the Nuclear Engine for Rocket Vehicle Application (NERVA) program. Canceled in 1969 because of high projected development costs, the NERVA program would have resulted in large heavy-lift launch vehicles capable of supplying a planned lunar base if successful. NERVA was also to have played an integral part in a proposed manned mission to Mars, which was at one time scheduled for launch in November 1981 with a manned landing in August 1982.

Rhyolite

First launch: 6 March 1973

The Rhyolite spacecraft were a series of four electromagnetic reconnaissance satellites placed in geosynchronous orbit by the US Air Force to intercept the telemetry originating during Soviet and Chinese missile launches. They were described as replacements for earlier Ferret satellites, though the details are still classified.

SAGE

(See Applications Explorer Missions)

SAMOS

Launch vehicles: Atlas Agena A (early launches)
Titan 3B/Agena D (later launches)
Launch: 11 October 1960 (SAMOS 1)
31 January 1961 (SAMOS 2)
9 September 1961 (SAMOS 3)
17 March 1972 (SAMOS 87)
Total weight: 4092 pounds (early spacecraft)
6600 pounds (later spacecraft)
Diameter: 58.5 inches

SAMOS, the Satellite and Missile Observation System, was a series of US Air Force spacecraft designed to conduct TV surveillance of the entire world from polar, nongeosynchronous orbit, transmitting pictures to earth during the part of its orbit that took it over US territory. The SAMOS spacecraft were also able to return film capsules. The launch of SAMOS 2 was the first successful launch, and the first operational launch was in 1963 after all military (DOD) launches came to be routinely classified. All the launches took place from Vandenberg AFB.

Satcom

(See RCA Satcom)

SBS

Launch vehicles: Delta 3910/PAM-D (SBS 1 and 2)
Space shuttle/PAM-D (SBS 3 and 4)
Launch: 15 November 1980 (SBS 1)
24 September 1981 (SBS 2)
11 November 1982 (SBS 3)
30 August 1984 (SBS 4)
Total weight: 2325.4 pounds (SBS 4, 7000 pounds)
Diameter: 85 inches
Height: 111 inches (260 inches with antennae deployed)
Systems: Repeater with ten 43 MHz channels, each with a 20-watt multicollector traveling wave tube, with the antenna gain developed by the single large-beam reflector

The SBS (Satellite Business System) spacecraft were designed to provide integrated, all-digital, interference-free transmission of telephone, computer, electronic mail and video teleconferenc-

The launch of SBS 4 from the space shuttle, 1984.

HUGHES
SBS
SBS 4
NASA

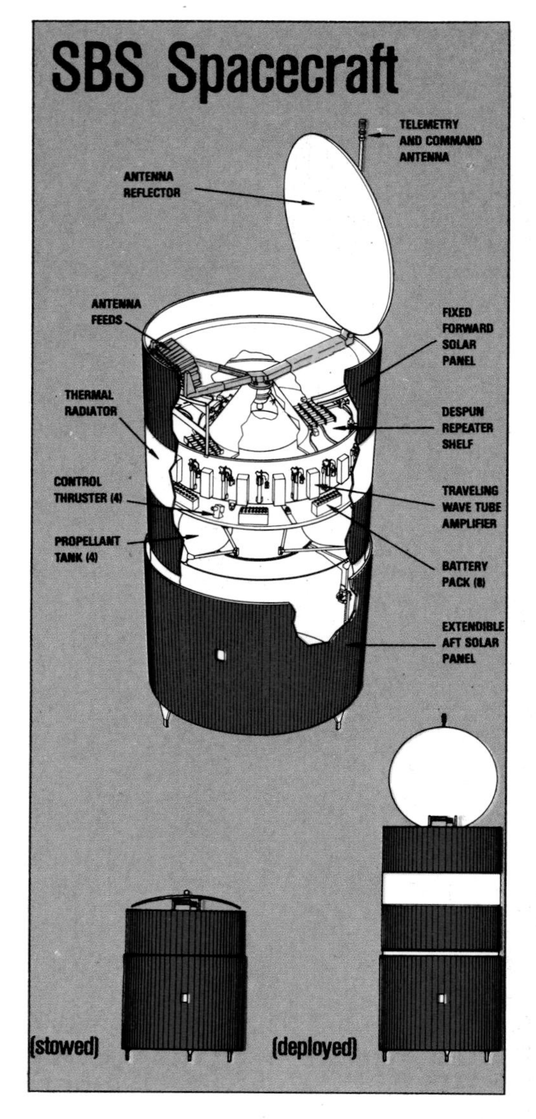

ing to business and industrial clients from geosynchronous earth orbit. The SBS domestic satellite system provided fully switched private networks to businesses, government agencies and other organizations with large and varied communication requirements. The spacecraft were built by Hughes and launched by NASA on a reimbursable basis for SBS, which was a consortium of Aetna Insurance, COMSAT General and IBM. Initial customers for the services of the SBS spacecraft included Boeing Computer Services, General Motors, Travelers Insurance Companies and Westinghouse Electric. The third SBS spacecraft (SBS 3) had the distinction of being the satellite deployed from the space shuttle by means of the payload assist module (PAM).

SCATHA

Launch vehicle: Delta 2914
Launch: 30 January 1979
Total weight: 1444 pounds at launch
787 pounds in orbit
Diameter: 6 feet
Height: 6 feet
Shape: Cylindrical
Systems: Charging electrical-effects analyzer, satellite surface potential monitor, electrical sheath fields, energetic proton detector, high-energy particle spectrometer, rapid-scan particle detector, thermal electronic measurement system, light ion-mass spectrometer, energetic ion spectrometer, particle detectors, electric-field detector, magnetic-field monitor, thermal control/ contamination experiments and particle beam systems

The external walls of the spacecraft were covered with solar cells to provide SCATHA with 300 watts of electrical power. The structure housed an apogee kick motor.

The objective of the SCATHA (Spacecraft Charging at High Altitudes) project was to investigate the electrical charging effects on spacecraft in high-altitude orbits caused by their passing through the earth's magnetic field. The spacecraft was built by Martin Marietta and launched by NASA from Cape Canaveral for the US Air Force Space Division.

SCORE

Launch vehicle: Atlas B
Launch: 18 December 1958
Last transmission: 31 December 1958
Re-entry: 21 January 1959
Total weight: 154 pounds

The US Army's SCORE (Signal Communication by Orbiting Relay Equipment) project involved the launch of the first-ever communications satellite. The SCORE spacecraft was launched from Cape Canaveral into an elliptical earth orbit from which it transmitted retaped messages for two weeks. Among the

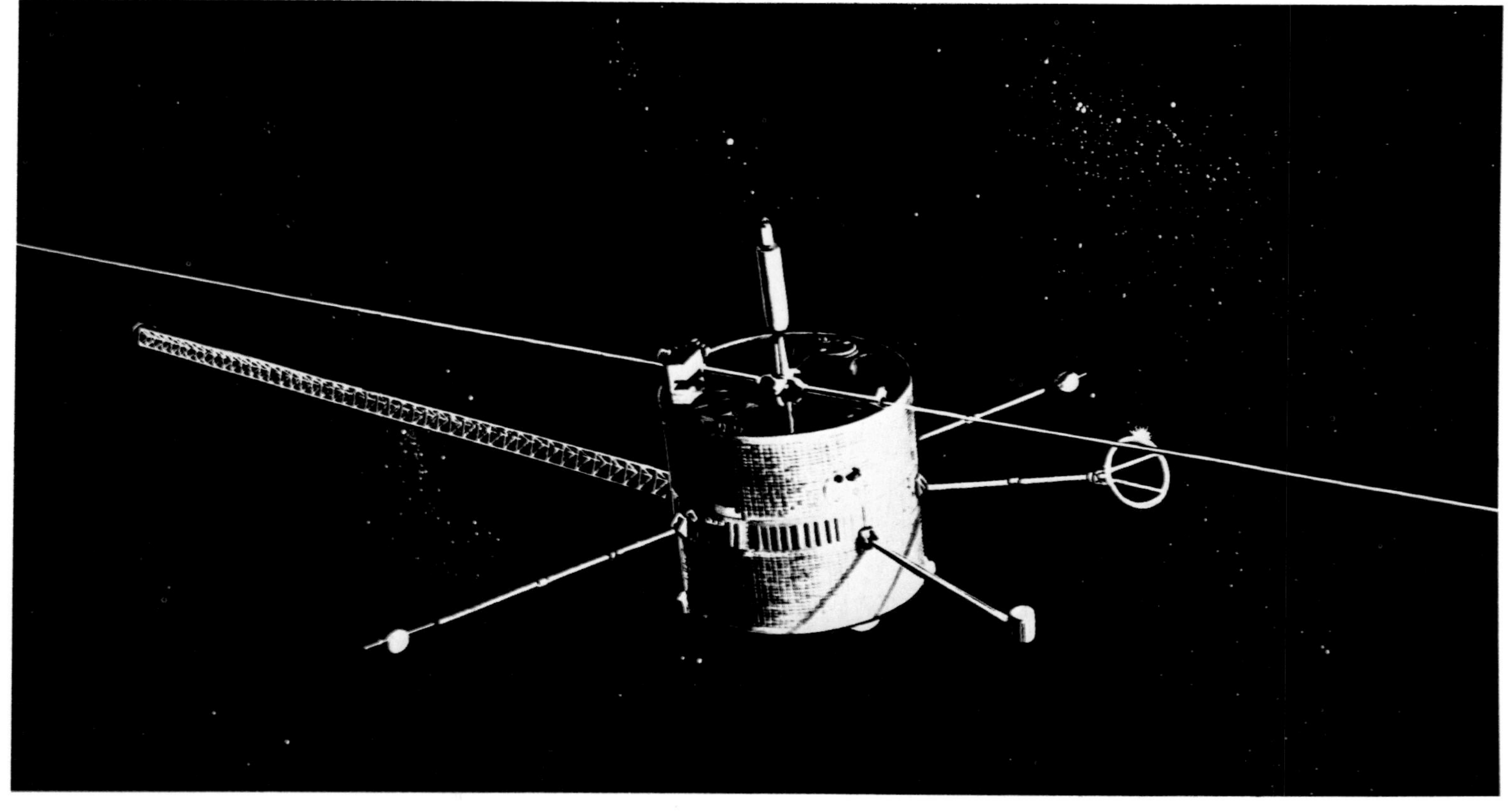

Below: **An artist's concept of the SCATHA on a host Project 78-2 USAF satellite.**

messages included was a Christmas greeting from President Eisenhower.

Seasat

Launch vehicle: Atlas F
Launch: 26 June 1978

The objective of the Seasat project was to demonstrate techniques for global monitoring of oceanographic phenomena and features, such as surface temperature, wave topology, surface wind speed and direction and ice-field dynamics. The difference between Seasat and Landsat was the use of active and passive microwave sensors, respectively, to achieve all-weather capacity. Launched from Vandenberg AFB, Seasat was managed for NASA by the Office of Space and Terrestrial Applications.
(See also Landsat)

Secor

Launch vehicle: Thor AD/Agena D
Launch: 18 May 1968 (Secor 10)
14 April 1969 (Secor 13)
Total weight: 45 pounds

The Secor spacecraft consisted of a rectangular package 9 by 11 by 13 inches, with eight antennae on the side and one on top. It carried a 12-pound solid-state transponder to retransmit signals from four ground stations, three at known positions and the fourth at the position to be determined by Secor.

Right: **Seasat failed after 1502 orbits.**
Below: **SCATHA satellite readied for launch.**

The US Army's Secor, or Sequential Collation of Range spacecraft, built by Cubic Corporation (with transponders by ITT), were designed as navigational satellites and part of the Army's overall Secor position-determination program. As a navigational system, Secor preceded the US Navy's Transit System and the much more ambitious USAF/DOD NAVSTAR Global Positioning System. Nevertheless, most of the Secor launches remain classified, except for two that were piggybacked aboard NASA's Nimbus.
(See also NAVSTAR, Transit)

SERT

The SERT (Space Electric Rocket Test) project was conceived to assess the performance of electron-bombardment ion engines over long periods of time and how such prolonged operations might affect other spacecraft systems. It had been proposed that such low-thrust, high-performance rocket engines be used in orbiting earth satellites or in deep-space missions to the outer planets and beyond.

SERT 1
Launch vehicle: Scout
Launch and re-entry: 20 July 1964

SERT 1's contract ion thruster could not be started because of a short circuit, but the electron-bombardment ion thruster was operated successfully for 31 minutes. It was the first use of an ion thruster in space and confirmed that high-prevalence ion beams could be neutralized in space

SERT 2
Launch vehicle: Thor Agena

Launch: 4 February 1970
Total weight: 3100 pounds
Diameter: 60 inches
Length: 248 inches

SERT 2 had two electron-bombardment ion engines generating 1 ounce (.006 pound) of thrust by electrically ionizing vaporized mercury (propellant), accelerating and neutralizing the ions, and expelling them at 50,000 mph. The first engine aboard SERT 2 was started on 14 February 1970 and gradually brought up to full power. It was shut down and the second engine was started and operated for 3785 hours (five months) until a short circuit caused it to shut down. Unable to restart it, NASA restarted the first engine on 24 July. This engine failed after 2011 hours and the project was designated unsuccessful. However, SERT did demonstrate the potential of ion propulsion and advance the technology.

Skylab Space Station

(See Appendix 1)

Small Astronomy Satellite

(See Explorer 53)

Small Scientific Satellite

(See Explorer 45)

Solar Explorer

(See Explorers 30, 37)

Solar Maximum Mission

Launch vehicle: Delta 3910
Launch: 14 February 1980
Total weight: 5093 pounds
Diameter: 7 feet
Height: 13 feet
Systems: Gamma ray spectrometer, hard X-ray spectrometer, hard X-ray imaging spectrometer, high-altitude observatory coronagraph/polarimeter, soft X-ray polychromator, solar (spectrum) constant monitor, ultraviolet polarmeter and ultraviolet spectrometer

The Solar Maximum Mission (SMM, or Solar Max) spacecraft was launched near the peak of a period of maximum solar-flare activity in an attempt to answer important scientific questions regarding solar flares and the radiation and particle beam emissions from them. The NASA-developed spacecraft functioned satisfactorily, observing solar flares and solar radiation until attitude control was lost in December 1980. The spacecraft was retrieved, however, by the manipulator arm of the space shuttle *Challenger* in April 1984 and successfully repaired by astronauts George Nelson and James van Hoften. It was re-released into space on 12 April to continue observations of the sun and later of Halley's Comet.

The upper portion of the spacecraft was the instrument module that housed all solar observation instruments and the fine-pointing sun-sensor system for aiming control. Below the instrument module was the multimission modular spacecraft (MMS), a 5-foot triangular framework that housed the essential attitude control, power, communications and data-handling systems. Two fixed solar paddles were attached to a transition adaptor between the upper instrument module and the lower spacecraft bus. The paddles supplied power to the spacecraft during the daylight portion of orbits, while three rechargeable batteries supplied power at night.

Solar Max (*above*) is mounted on the MSS (*left*). Launched in 1980, it was recovered in April 1984 by the space shuttle (*right*).

Solar Mesosphere Explorer

Launch vehicle: Delta
Launch: 6 October 1981
Diameter: 50 inches (satellite body)
Height: 69 inches

Systems: Ultraviolet ozone spectrometer, a 1.27-micron spectrometer, nitrogen dioxide spectrometer, four-channel infrared radiometer, solar ultraviolet monitor

The Solar Mesosphere Explorer (SME) spacecraft was an atmospheric research satellite designed to study reactions between sunlight, ozone, and other chemicals in the atmosphere and how concentrations are transported in the atmosphere. The spacecraft was developed by Ball Aerospace and the scientific systems by the University of Chicago Laboratory for Atmospheric and Space Physics.

The spacecraft was powered by a disk-shaped solar array 88 inches in diameter. The solar array was fixed to the lower end of main cylinder and charged nickel-cadmium batteries.

SOLRAD

(See Explorer 44)

Typical Solar Mesosphere Explorer (SME) Ground Tracks

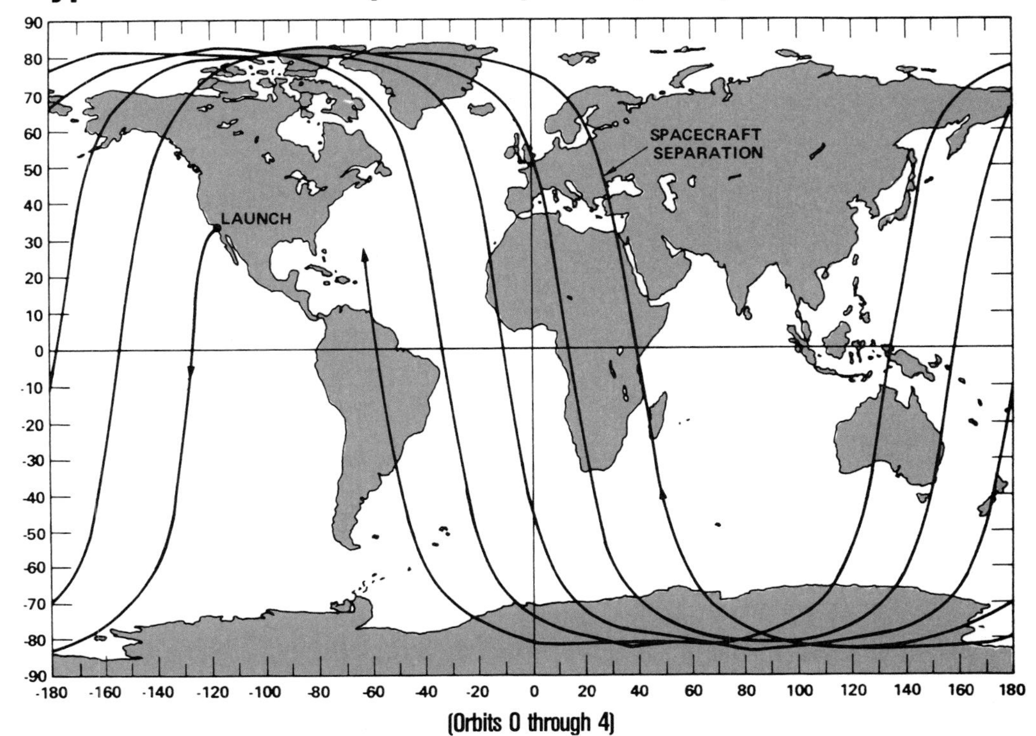

(Orbits 0 through 4)

Space Shuttle

Officially known as the Space Transportation System (STS), the space shuttle program is the most important earth-orbit space program ever undertaken by the United States, and ranks along with Apollo as one of the two most important US space programs ever. Very simply, the STS consists of an orbiting vehicle (OV), or orbiter, which is launched into space by a specially designed launch vehicle in the same way that spacecraft are traditionally launched. (The vehicles are also referred to by the name space shuttle.) However, upon re-entering the earth's atmosphere, the delta-winged orbiter can glide up to 1100 miles under aerodynamic control to an aircraftlike landing and then be reused. The earlier manned spaceflights all used space capsules that were not designed for reuse. The STS can be used to carry scientific experiments that can be successfully completed only in the weightlessness of space, and it can be used to carry satellites and place them into orbit at a much lower cost than an expensive one-time launch vehicle. It can be used to retrieve satellites from orbit or to repair them in space. Supported by both NASA and DOD, the shuttle can carry both civilian and military payloads. All the early shuttle launches were intended to take place from the Kennedy Space Center at Cape Canaveral, and a second (primarily military) launch facility was established at Vandenberg AFB in California for use commencing in 1985. The typical STS mission profile involves a duration of seven days, but the system is designed with a capability for missions of up to 30 days. The original idea for the present space shuttle dates back to 1969 when the Presidential/NASA/DOD Space Task Group was asked to come up with an idea for the next generation of American manned spacecraft. The basic idea, for a reusable 'space plane' rather than an expendable space capsule, was not new. In Germany Dr Eugen Sanger had developed a reasonably practical concept for such a machine in the early 1940s, and both NASA and the US Air Force had conducted tests with high-altitude research rocket planes in the 1950s and 1960s. The most important of these was the X-15 rocket plane, which made 199 flights between 1959 and 1963, reaching altitudes of 67 miles. It was built by North American Aviation, the company that later became Rockwell International, the builder of the space shuttle. During this period, Boeing and NASA developed a concept for a system that was very much like the ultimate STS. The Dynamic Ascent and Soaring Flight (Dyna Soar) program resulted in a rocket plane designated X-20 that could be launched into space and return to earth like an airplane. The progam was canceled in 1962 without construction of a prototype, but the concept survived and had a great deal of influence on the early conceptual development of the space shuttle a decade later.

***Above: Columbia* (OV-102) STS-1 with the white tank thunders into space. *Columbia* with the unpainted tank (*right*) on mission STS-3.**
***Below:* OV-101 *Enterprise* during initial flight testing in August 1977.**

The final design for the space shuttle was approved in March 1972, and Rockwell was given the go-ahead to begin construction on the spacecraft portion of the system. It was to be America's largest spacecraft and would have a specially developed launch vehicle system consisting of two enormous Martin Marietta solid rocket boosters (SRB) and an even larger external fuel tank. The latter were designed to parachute into the ocean and be preserved for later use.

The first orbiter was completed by Rockwell's Palmdale, California facility and delivered to Edwards AFB, where testing began in February 1977. This first orbiter was designated OV-101 (Orbiter Vehicle, first) and nicknamed *Enterprise* after the Hollywood starship in the television series and movie *Star Trek*. *Enterprise* was never intended for actual spaceflight, but for tests that would be conducted in earth atmosphere. A Boeing 747 transport was specially modified to carry the orbiter on its back during the initial tests because the orbiter was not designed to take off in level flight. The 747 was also used later in the program to transport the orbiters from place to place within the atmosphere.

The first free flight of the *Enterprise* came on 12 August 1977, with Gordon Fullerton and Fred Haise at the controls. *Enterprise* was released from the 747 at 22,800 feet, whereupon the crew glided the huge spacecraft prototype to a perfect landing. In May 1979, after two months of successful flight testing and eight months of structural tests, *Enterprise* was mated to the shuttle launch vehicle system at Kennedy Space Center for further tests. Meanwhile, in March 1979, the orbiter *Columbia* (OV-102), the first spaceflight-rated orbiter, arrived at KSC. After two years of extensive tests and revisions to the engine and heat shield, *Columbia* made its first flight on 12 April 1981, three years and one month behind the schedule set in 1972 when the program began.

SA

The STS Orbiting Vehicle

The reusable orbiting vehicle is manufactured by the Space Division of Rockwell International. It has a total payload capacity in excess of 32,000 pounds (the weight of the ESA Spacelab) and can deliver single or multiple payloads into orbit from its 15-by-60-foot payload bay. The orbiter is roughly the size of a midsize commercial airliner. It is constructed of an aluminum framework, and its exterior is covered with thermal protective materials to shield the spacecraft from solar radiation and the extreme heat of atmospheric re-entry. The top of the orbiter is covered with coated silica and the sides with coated flexible sheets that can protect the ship at temperatures up to 1200 and 700 degrees Fahrenheit respectively. The bottom of the orbiter and the leading edge of the tail are covered with glossy black silica tiles that protect them at temperatures up to 2300 degrees Fahrenheit. A black reinforced carbon material covers the nose and wing leading edges for protection at the same temperatures. Beginning with the orbiter *Discovery* (OV-103), many of the white tiles were replaced with fibrous refractory composite insulation (FRCI) or silicon fiber blankets, which have resulted in a half ton of weight savings.

The 15-by-60-foot payload bay can accommodate a wide variety of cargo, scientific experiments and satellites. The payload(s) in the bay can be moved, operated and deployed by either the Canadian-built Spar Aerospace Ltd remote manipulator arm or by payload specialists in space suits. The Spacelab module, a 9-by-22-foot-6-inch manned space laboratory designed and built in Europe under the auspices of the European Space Agency, was designed to fit into the STS payload bay and was carried for the first time in November 1983. Spacelab is connected directly to the orbiter's cabin and is pressurized to the same level so that payload specialists can work there in a 'shirtsleeves' environment without space suits. The Spacelab's total weight is 32,000 pounds.

Forward of the payload bay is the cabin, providing living and working as well as storage space for the crew. It is pressurized to a pressure of 14.7 pounds per square inch (psi) to simulate sea level, whereas the earlier manned US spacecraft were pressurized to 5 pounds psi. The atmosphere is a nitrogen/oxygen mixture like that of earth, unlike the pure oxygen atmosphere of earlier US manned spacecraft. The cabin's upper level contains the flight deck and controls for all phases of

The Orbiting Vehicle Spacecraft of the Space Transportation System (STS)

Orbiting Vehicle Designation	Orbiting Vehicle Name	Weight Unloaded (dry)	First Flight*
OV-101	*Enterprise*	149,600 lb	12 August 1977*
OV-102	*Columbia*	152,128 lb	12 April 1981
OV-99**	*Challenger*	149,642 lb	4 April 1983
OV-103	*Discovery*	147,980 lb	30 August 1984
OV-104	*Atlantis*	147,980 lb***	20 September 1985

*First flights are flights into space except for *Enterprise*, which was not intended for spaceflight
***Challenger*, like *Enterprise*, was built as a nonspaceflight test vehicle, but was later modified for spaceflight
***Estimated

General Specifications for All Orbiting Vehicles

Length	Wingspan	Tail Height
122 feet 2.5 inches	78 feet 0.07 inch	56 feet 7 inches

Top: Astronaut Anna Fisher operated the *Discovery* RMS to retrieve two stranded satellites in November 1984. *Above: Columbia* lands safely at Edwards AFB. *Right: Challenger* (STS-7) with payload pallets and cradles exposed.

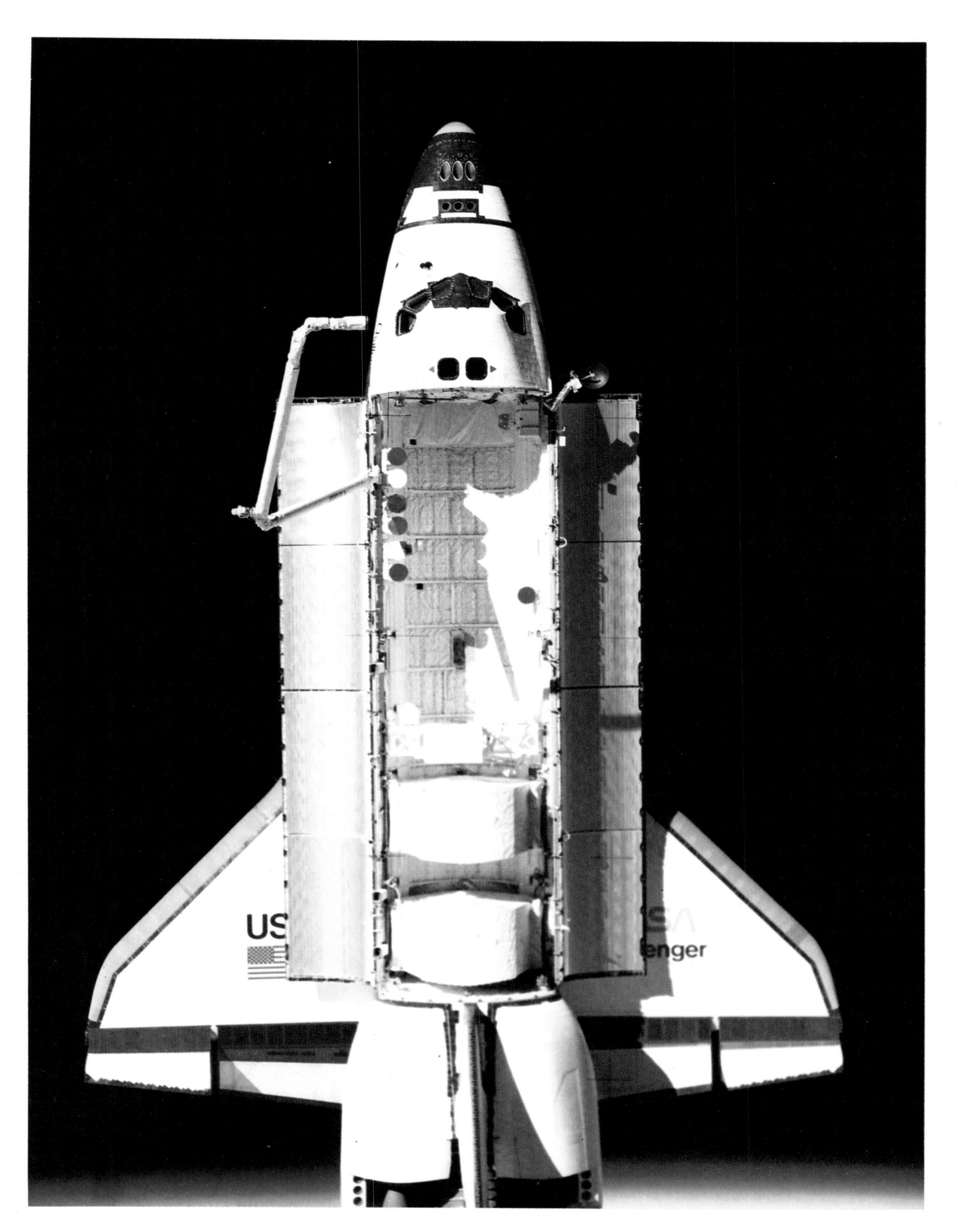
US
enger

the spacecraft operations, including those directed to the spacecraft's payload. The lower deck contains the galley, eating area, sleeping area and zero-gravity bathrooms.

The basic crew includes a pilot and orbiter commander who sit in a traditional pilot-copilot seating arrangement on the flight deck, and the mission specialist, who is a scientist or technician rather than a pilot. The earlier missions were two-man flights, flown only by pilot and commander. The flight-deck controls and other systems are arranged to permit the orbiter to be landed by a single crew member. In addition to these three crew members, there are provisions for up to four technical personnel or payload specialists to manage the operation or deployment of the payload. The first time that a mission carried the maximum seven crew members was in October 1984 aboard the orbiter *Challenger* (OV-99). In the case of a rescue mission, however, the orbiter could carry three additional passengers by a reconfiguration of the lower-level sleeping quarters. All the shuttles carry provisions for extravehicular activity, or spacewalks, for two crew members. EVAs take place on missions where access to the payload bay and the payloads is necessary. The payload specialists exit through an airlock and hatch in the lower level of the cabin.

The orbiter's three main engines, located in the aft fuselage below the tail in a triangular pattern, are manufactured by the Rocketdyne Division of Rockwell International. Each delivers between 375,000 (sea-level) and 470,000 (high-altitude) pounds of thrust, and these liquid-fuel engines are capable of 7.5 hours of continuous use without maintenance or overhaul. Given that the engines are used on average for only eight minutes per mission, the engines can be used for 55 separate missions, unless especially heavy payloads are being lifted into orbit.

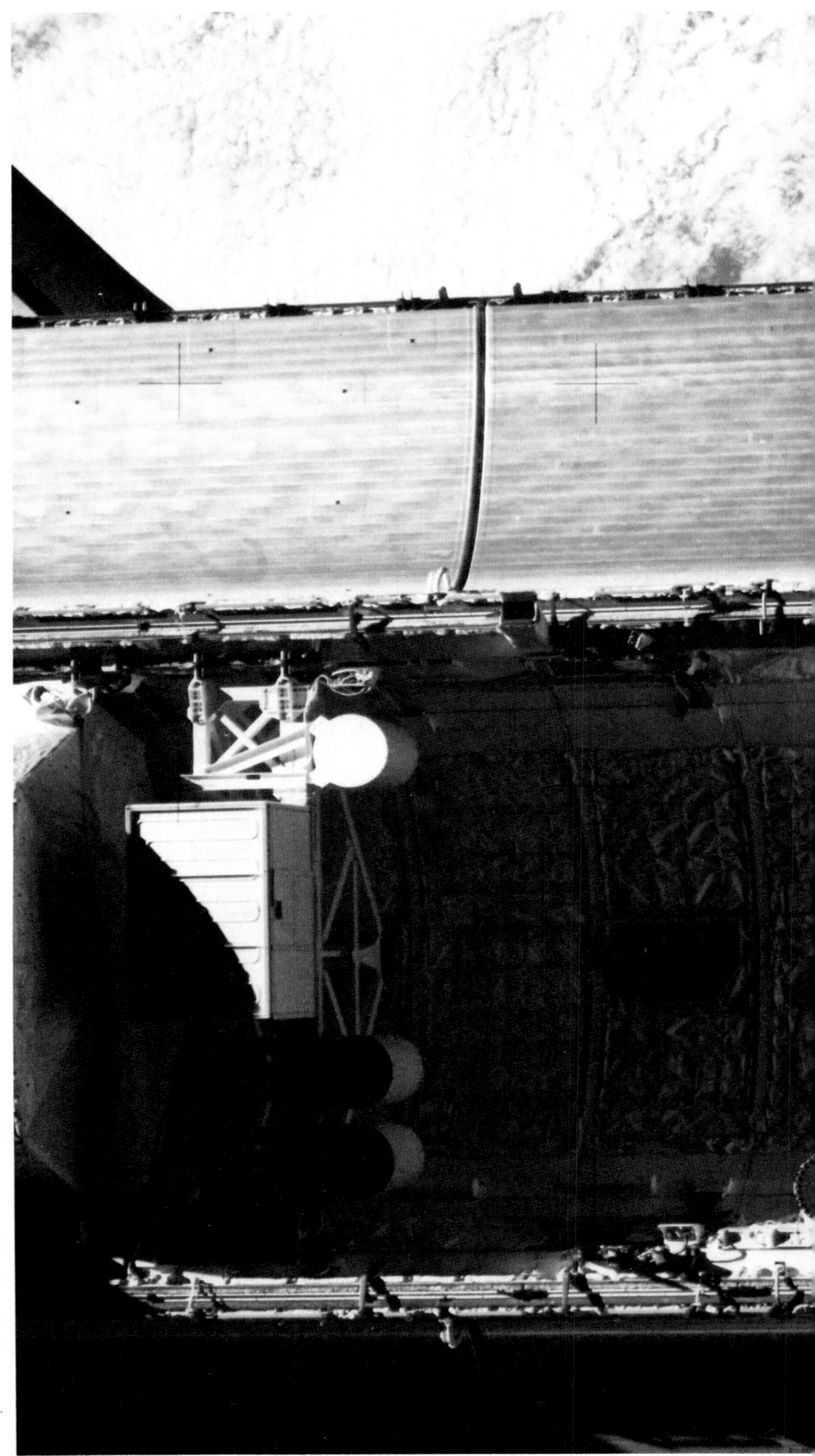

Above: **Story Musgrave during his STS-6 EVA.** ***Right:*** **A satellite view of the orbiter *Challenger* during STS-7, showing the open cargo bay and the remote manipulator arm.**

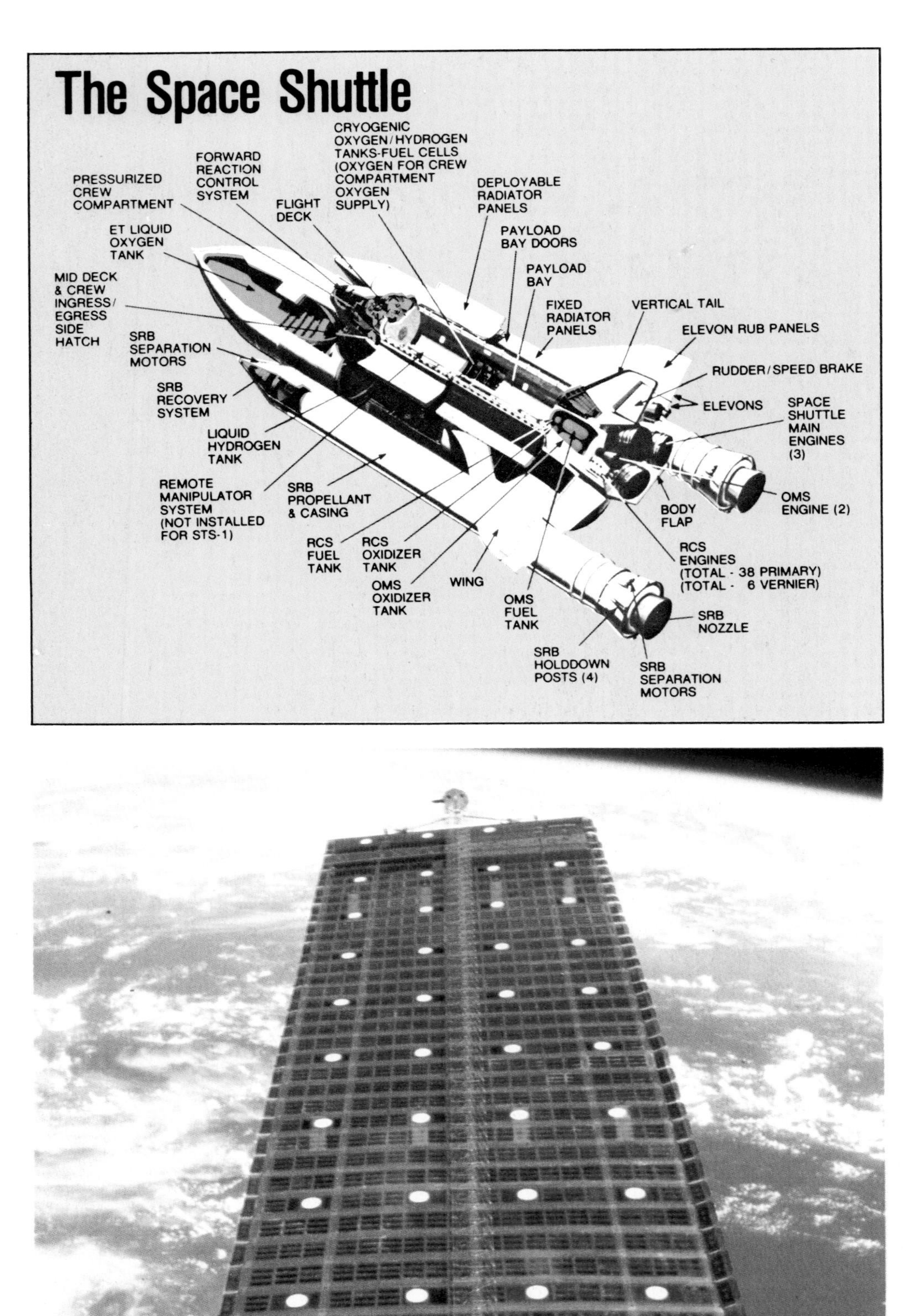

Above: Extended solar array experiment panel exposed by the mission 41-D crew members.

Right: Dale Gardner, tethered to *Discovery*'s starboard side during 14 November 1984 EVA.

STS Launch Vehicle System

The STS launch vehicle is covered in specific detail in the appendix, but because it was specially designed and integrated into the overall STS, a few details here are in order. The two solid-fuel rocket boosters used for each flight are the largest solid-fuel rockets ever flown and the first ever designed for reuse. They are used to lift the orbiter to an altitude of 27.5 miles, when the orbiter's engines take over. At this point, eight separation motors separate the boosters and they drop into the ocean 160 miles down range of launch, assisted first by a pilot parachute, then a drogue chute, and finally by three 104-foot-diameter main parachutes.

The largest item in the STS is the huge external tank. It is designed to serve as the 'structural backbone' of the system during launch and as a container for the propellants. The pointed, top part of the tank contains 666 tons of liquid-oxygen propellant, which is delivered to the orbiter at a rate of 16,800 gallons per minute. The larger portion of the tank is filled with liquid-hydrogen propellant weighing 112 tons and delivered at a rate of 45,283 gallons per minute. The tank is released just prior to the orbiter's achieving full orbital velocity and drops into the ocean. It is the only major part of the STS that is not reused.

The tank was developed by Martin Marietta and is built at the Michoud Assembly Facility in New Orleans, where the first stages of the huge Saturn 1B and Saturn 5 launch vehicles of the Apollo and Skylab programs were built. In the early STS flights, the external tank was painted white like other orbiters and boosters, but the use of a simple rust-colored primer in later missions has saved a half ton of weight.

STS Flight Designations

The designation system for the flights of the STS program began with the first STS flight into space in 1981, which was appropriately designated STS-1. This was the first flight of the orbiter *Columbia*, the *Enterprise* flights not having been designated because they were just glide tests within the earth's atmosphere. The STS designation series proceeded in chronological order, irrespective of which orbiter was used. For example, STS-5 with *Columbia* was followed by STS-6 with *Challenger*. The STS series continued through the end of fiscal year (FY) 1983 and the flight of STS-8. This followed the first flight of FY 1984, which carried the dual designations STS-9 and 41A and was the last flight to carry an STS designation.

The new three-digit designator adopted at the beginning of FY 1984 is a bit more complex, but it also carries more information about an individual flight. The first digit of the code is the last digit of the fiscal year in which the flight was *initially scheduled*, thought not necessarily in the year in which it actually took place. In the case of 41A, the designation shows a flight that was scheduled for FY 1984.

The second digit indicates the launch site of the mission. The numeral 1 stands for the Kennedy Space Center at Cape Canaveral, and the numeral 2 stands for Vandenberg AFB, California. The numeral 3 will be assigned to a third STS

Left: **41-D crew commander Henry Hartsfield, Jr loads the IMAX camera, first used on mission 41-C.** ***Right:*** **Bruce McCandless first used the MMU on mission 41-B.**

The First Seventeen Space Shuttle Transportation System (STS) Missions

Number	Mission Designation	Orbiting Vehicle	Launch Date	Duration in Days	Crew Members (number in parentheses)	Results
1981						
1	STS-1	*Columbia*	12 Apr	2	(2) Robert Crippen, John Young	First STS flight; first use of solid rockets in manned spaceflight; first US spacecraft to land on land.
2	STS-2	*Columbia*	12 Nov	2	(2) Joe Engle, Richard Truly	First test of remote manipulator arm; carried OSTA 1 earth resources pallet with SIR-A radar.
1982						
3	STS-3	*Columbia*	22 Mar	8	(2) Gordon Fullerton, Jack Lousma	Carried OSS-1 space science experiments.
4	STS-4	*Columbia*	27 June	7	(2) Henry Hartsfield, Thomas Mattingly	First hard-surface landing.
5	STS-5	*Columbia*	11 Nov	5	(4) Joseph Allen, Vance Brand, William Lenoir, Robert Overmeyer	First operational STS mission; first to launch satellites; launched SBS-3 (US) and Anik-C (Canada).
1983						
6	STS-6	*Challenger*	4 Apr	5	(4) Karol Bobko, Story Musgrave, Donald Peterson, Paul Weitz	First flight of *Challenger*; first EVA from an STS orbiter (Musgrave, Peterson).
7	STS-7	*Challenger*	18 June	6	(5) Robert Crippen, John Fabian, Frederick Hauck, Sally Ride, Norman Thagard	First US woman in space (Ride); first person to fly in STS twice (Crippen); first use of remote manipulator arm to deploy a satellite, a German SPAS.
8	STS-8	*Challenger*	30 Aug	6	(5) Guion Bluford, Daniel Brandenstein, Dale Gardner, William Thornton, Richard Truly	First black US astronaut (Bluford); first night launch; first use of payload flight test article.
9	STS-9 (41-A)	*Columbia*	28 Nov	10	(6) Owen Garriott, Byron Lichtenberg, Ulf Merbold, Robert Parker, Brewster Shaw, John Young	First flight of European (ESA) Spacelab module; first non-American astronaut on a US spaceflight (Merbold, Germany).
1984						
10	41-B	*Challenger*	3 Feb	8	(5) Vance Brand, Robert Gibson, Bruce McCandless, Ronald McNair, Robert Stewart	First use of manned maneuvering unit (MMU) for EVA (McCandless, Stewart). Launched 3 satellites: SPAS-01A (Germany), Palapa B-2 (Indonesia) and Westar 6 (US); the latter 2 launches failed to place spacecraft in proper orbit due to payload assist module (PAM).

launch facility, should one be built anywhere in the world.

The final digit is a sequence indicator for a specific launch site. For example, 61B stands for the second launch from KSC scheduled during FY 1986, and it can exist alongside 62B, which indicates the second launch scheduled from Vandenberg AFB during the same fiscal year. The sequence letters need not all be used, however. For example, in 1984 mission 41D was originally scheduled for 22 June but had to be postponed until 30 August, the date set for 41G. Thus, the planned 41E and 41F missions scheduled for the summer of 1984 were deleted from the roster. Mission 41G was then moved ahead to 5 October and mission 41H was canceled, making 41G the last flight of that fiscal year (see chart below).

The First Seventeen Space Shuttle Transportation System (STS) Missions

Number	Mission Designation	Orbiting Vehicle	Launch Date	Duration in Days	Crew Members (number in parentheses)	Results
1984						
11	41-C	*Challenger*	6 Apr	7	(5) Robert Crippen, Terry Hart, George Nelson, Francis Scobee, James van Hoften	Repaired Solar Maximum Mission spacecraft; longest EVA (7 hr 18 min) since Apollo 17; deployment of a Long Duration Exposure Facility, the largest payload yet deployed by STS.
12	41-D	*Discovery*	30 Aug	5	(6) Michael Coats, Henry Hartsfield, Steven Hawley, Richard Mullane, Judith Resnik, Charles Walker	First astronaut representing private industry (Walker) operated McDonnell Douglas Electrophoresis project to explore pharmaceutical processing in space; launched Leasat 2, Telstar 3, SBS 4; deployed 102-foot OAST 1 solar array.
13	41-G	*Challenger*	5 Oct	8	(7) Robert Crippen, Marc Garneau, David Leestma, Jon MacBride, Sally Ride, Paul Scully-Power, Kathryn Sullivan	First Canadian in space (Garneau); first US EVA by a woman (Sullivan); deployment of ERBS satellite first oceanographer to document ocean phenomena from space (Scully-Power); use of large-format camera to provide high resolution terrain imagery; demonstration of satellite refueling.
14	51-A	*Discovery*	8 Nov	9	(5) Joseph Allen, Anna Fisher, Dale Gardner, Frederick Hauck, David Walker	First space shuttle flight to retrieve wayward satellites (Palapa B-2 and Westar 6); also launched Leasat and Telesat satellites.
1985						
15	51-C	*Discovery*	24 Jan	3	(4) James Buchli, Ellison Onizuka, Gary Payton, Loren Shriver	First semiclassified Defense Department launch. The all-military crew released a classified reconnaissance satellite into geosynchronous orbit through use of the IUS.
16	51-D	*Discovery*	12 Apr	5	(7) Karol Bobko, E J 'Jake' Garn, David Griggs, Jeffrey Hoffman, Rhea Seddon, Charles Walker, Donald Williams	Deployed Syncom 4 and Telesat satellites; Experiments included Electrophoresis and Echo-Cardiograph. First US Senator in Space (Garn).
17	51-B	*Challenger*	29 Apr	7	(7) Frederick Gregory, Don Lind, Robert Overmyer, Norman Thalgard, William Thorton, Lodewijk van den Berg, Taylor Wang	Ten experiments conducted in ESA Spacelab 2; First Dutch astronaut (van den Berg).

Squanto Terror

Launch vehicle: Thor
First launch: March 1964

With its particularly fearsome code name, the Squanto Terror program (Air Force Project 437), was the first known serious attempt by the US Air Force to develop a practical antisatellite capability. The system involved Thor-launched spacecraft carrying nuclear explosives that could knock out enemy spacecraft by direct contact or explosive concussion. They were loosely based on an October 1962 USAF/AEC test, code named Starfish, in which a nuclear-weapon test was conducted in outer space. In the Starfish test it was noted that high-energy particles, driven at high speed by the nuclear explosion, damaged or destroyed the systems of many satellites in the vicinity.

The early Squanto Terror tests were extremely successful, with dummy warheads coming within a mile of their targets. By 1965, after only three tests, Squanto Terror was declared operational. A total of 16 test launches, all from Johnson Island in the Pacific, were conducted by the time the tests concluded in 1968. Meanwhile a 1967 treaty was signed that banned nuclear weapons from space, and by 1975 the Johnson Island launch facility had been put out of service.

Surveyor

The object of the Surveyor program was to soft-land a series of spacecraft on the surface of the moon to demonstrate the feasibility of such a maneuver in advance of the projected Apollo manned lunar landings. Once this was determined, the Surveyor landers were used to help evaluate potential Apollo landing sites. The program was also intended to demonstrate the capability of spacecraft to make the necessary midcourse flight-path corrections required to take it to the moon. The project was undertaken in 1961 by NASA's Jet Propulsion Laboratory, and the Surveyor spacecraft were designed and built by the Hughes Aircraft Company.

Surveyor 1
Launch vehicle: Atlas Centaur
Launch: 30 May 1966
Lunar landing: 2 June 1966
Launch weight: 2194 pounds
Landing weight: 596 pounds
Diameter: 14 feet (across extended landing legs)
Height: 10 feet
Shell composition: Aluminum
Photos returned: 11,237 (10,338 during first lunar day, or first 12 earth days)

A triangular structure provides mounting surfaces and attachments for landing gear, the main retrorocket engine, vernier engines, and associated tanks, thermal compartments and other assemblies. Planar high-gain antennae and a solar panel, supported by a positioner movable along four axes, are mounted on a mast that extends from the top. A crushable aluminum footpad was attached to the end of each leg of the tripod landing gear, hinged to each of the three lower corners of the frame and folded into the nose shroud during launch. Two omnidirectional conical antennae were mounted on the ends of folding booms hinged to the spaceframe.

Surveyor made a successful soft landing in one inch of dust in the Ocean of Storms.

Surveyor 2
Launch vehicle: Atlas Centaur
Launch: 20 September 1966
Lunar landing: 23 September 1966
Launch weight: 2204 pounds
Landing weight: 644 pounds
Specifications and description: Same as Surveyor 1
Photos returned: None

Only two of three rockets were fired during the midcourse correction, putting the Surveyor 2 spacecraft into a tumble from which it never recovered. It crashed in Sinus Medii on the moon.

Surveyor 3
Launch vehicle: Atlas Centaur
Launch: 17 April 1967
Lunar landing: 20 April 1967
Launch weight: 2283 pounds
Landing weight: 625 pounds
Specifications and description: Same as Surveyor 1
Photos returned: 6315

Highly reflective rocks confused the spacecraft's descent radar so that the engines did not cut off at an altitude of 14 feet as intended. As a result Surveyor bounced twice, first to 35 feet, then to 11 feet, before coming to rest. Surveyor 3 was the first of the series to be equipped with a surface-sampling device, a scoop

***Above:* Surveyor 7 landed near Crater Tycho.**

mounted on a flexible motor-driven arm. The sampling device dug four trenches of depths ranging up to 7 inches, and deposited the sample material in front of the spacecraft's television cameras. Surveyor 3 was shut down for the cold lunar night on 3 May and could not be revived afterward.

Surveyor 4
Launch vehicle: Atlas Agena
Launch: 14 July 1967
Lunar landing: 17 July 1967
Launch weight: 2290 pounds
Landing weight: 625 pounds
Specifications and description: Same as Surveyor 1
Photos returned: None

Surveyor 4 made a perfect flight until contact was lost 2.5 minutes before touchdown.

Surveyor 5
Launch vehicle: Atlas Agena
Launch: 8 September 1967
Lunar landing: 11 September 1967
Launch weight: 2290 pounds
Landing weight: 625 pounds
Specifications and description: Same as earlier Surveyors, but Surveyor 5 carried no surface sampling device
Photos returned: 18,006 on the first lunar day and 1043 on the second

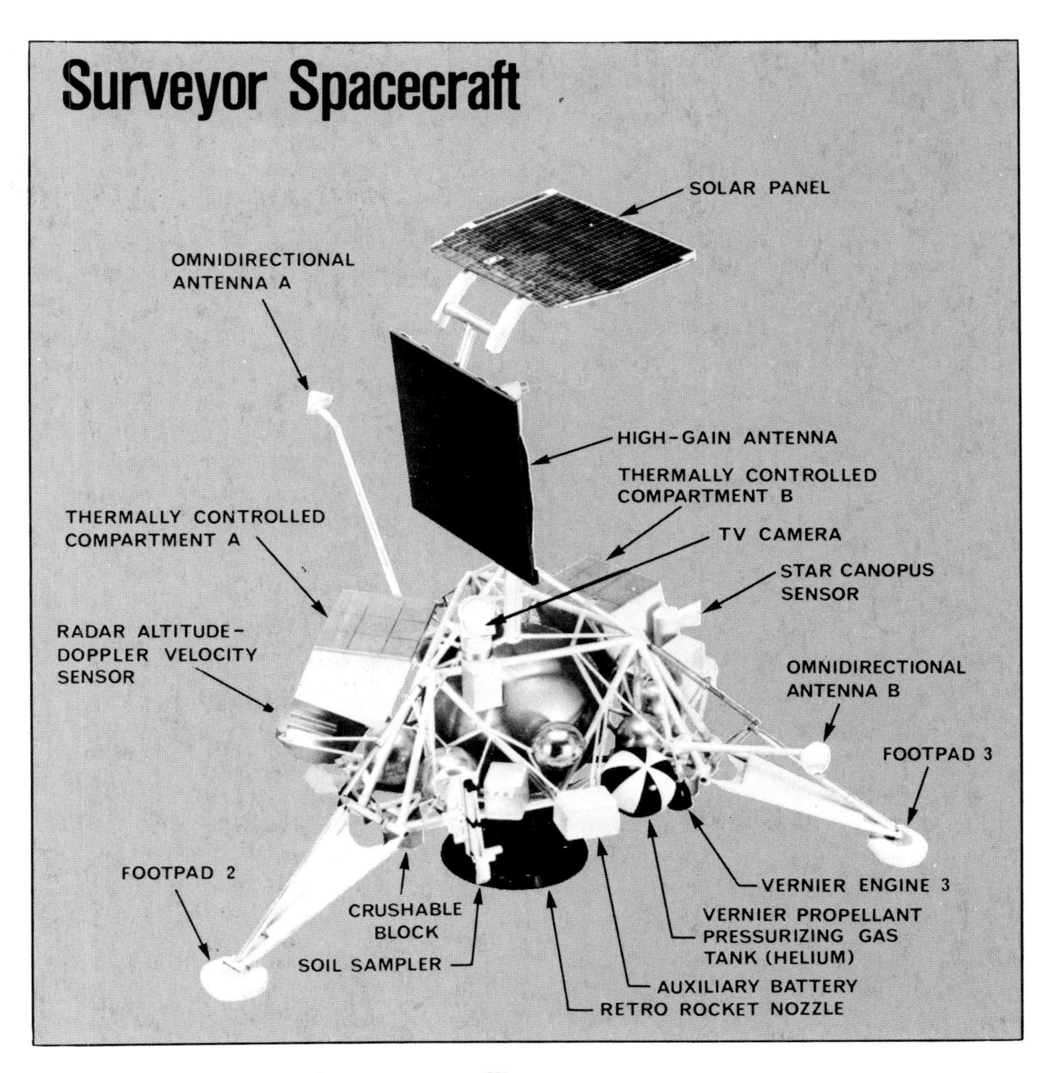

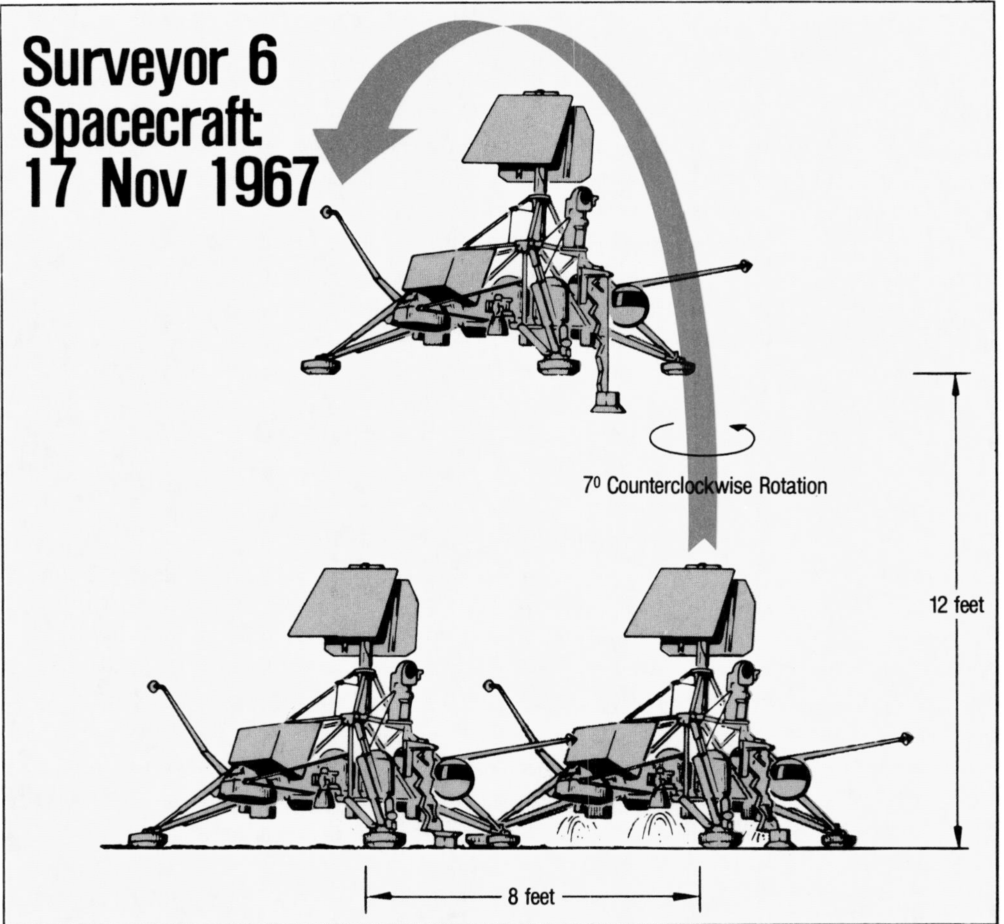

The mission nearly ended in failure because of a helium leak, but a risky nonstandard landing was successful and Surveyor 5 returned data for two weeks. During the first lunar day the spacecraft was shut down for the night and successfully restarted during the following lunar day, although the quality of the photos was lower. An alpha backscattering device (a miniature chemical laboratory) was carried in place of the surface-sampling scoop, and it returned data about the lunar surface soil indicating that it was composed of basaltic rock.

Surveyor 6
Launch vehicle: Atlas Centaur
Launch: 7 November 1967
Lunar landing: 10 November 1967
Launch weight: 2290 pounds
Landing weight: 625 pounds
Specifications and description: Same as Surveyor 5
Photos returned: 30,027

The spacecraft landed successfully and conducted both photographic and alpha backscattering soil surveys of the lunar environment. On 17 November, Surveyor 7 became the first spacecraft to lift off

from the lunar surface. The lander's engines were started and burned for 2.5 seconds, generating 150 pounds of thrust and lifting the spacecraft 12 feet above the moon's surface. Surveyor 6 was then moved 8 feet in a westerly direction, where it soft-landed and continued to function normally.

Surveyor 7
Launch vehicle: Atlas Centaur
Launch: 7 January 1968
Lunar landing: 10 January 1968
Launch weight: 2288 pounds
Landing weight: 625 pounds
Specifications and description: Same as Surveyor 1, but with both a surface sampling device *and* an alpha backscatter experiment
Photos returned: 21,046 during the first lunar day, 45 during the second.

The final spacecraft in the Surveyor series made a perfect landing less than two miles from its designated landing point. Both surface data experiments were carried and when the alpha instrument was not properly deployed, the surface sampler claw was used to push it down. All of the systems worked perfectly during the first lunar day, with a quarter of all the Surveyor series photographs taken by Surveyor 7 during this period. These television pictures included laser beams aimed from the earth to the moon. During the first lunar night, temperatures of −250 degrees Fahrenheit damaged the battery and only intermittent data was transmitted during the following day. Contact was lost on 20 February 1968.

Synchronous Meteorological Satellites

The Synchronous Meteorological Satellites, or SMS series spacecraft, were developed by Philco-Ford and managed by NASA for NOAA. They were geosynchronous orbit spacecraft and were direct predecessors of the GOES weather satellites.
(See also GOES)

SMS 1
Launch vehicle: Delta
Launch: 17 May 1974
Total weight: 1385 pounds at launch 630 pounds in orbit
Diameter: 75 inches

This view shows the unmanned Surveyor 3 (foreground) and the Apollo 12 LM on the moon.

SMS 1 is prepared for final systems test.

Height: 103 inches (including magnetometer)
Shape: Cylindrical

SMS 1 was spin stabilized and rotated at 100 rpm. Power was provided by 15,000 solar cells on six solar panels on the exterior of the drum and two 20-cell nickel-cadmium batteries.

The first weather observer in a geosynchronous orbit above the equator, SMS 1 gathered dust for the Global Atmosphere Research Program.

SMS 2
Launch vehicle: Delta
Launch: 6 February 1975
Total weight: Same as SMS 1
Diameter: 74.8 inches
Height: 136 inches (including magnetometer)
Shape: Drum

The SMS 2 consisted of an aluminum honeycomb shelf, supported by struts and an apogee boost motor. The experiment section was enclosed in a metallic cover and solar cell-covered side panels.

SMS 2 joined SMS 1 to provide a 24-hour weather watch on the Western Hemisphere and supplied cloud-cover photos every 30 minutes.

SMS 3
(See GOES 1)

Syncom

The Syncom series spacecraft were the first communications satellites ever placed in geosynchronous orbit. These experimental spacecraft were produced by Hughes Aircraft and operated by NASA to demonstrate the feasibility of geosynchronous communications satellites. They were the forerunners of the later Hughes communications spacecraft such as Intelsat.

A Hughes technician adjusts Syncom.

Syncom 1
Launch vehicle: Thor Delta
Launch: 14 February 1963
Total weight: 86 pounds
Diameter: 28 inches
Height: 15.5 inches
Shell composition: 3840 solar cells
Shape: Cylindrical

Syncom 1 was successfully placed in orbit, but radio contact was lost.

Syncom 2
Launch vehicle: Thor Delta
Launch: 26 July 1963
Specifications and description: Same as Syncom 1

Successfully placed in orbit over the Atlantic, Syncom 2 began regular service as planned on 16 August, which included a telephone conversation between President Kennedy of the US and Prime Minister Belewa of Nigeria.

Syncom 3
Launch vehicle: Thor Delta
Launch: 19 August 1964
Turned over to DOD: 1 January 1965
Specifications and description: Same as Syncom 1

Successfully placed in orbit over the Pacific, Syncom 3 began regular service on 23 September that included television relay of the 1964 Tokyo Olympics and teletype transmissions to aircraft on the Honolulu-San Francisco route.

Syncom 4
(See Leasat)

Tacsat

Launch vehicle: Titan 3C
First launch: 9 February 1969
Total weight: 1600 pounds at launch
1424 pounds in orbit
Diameter: 111 inches
Height: 300 inches
Shape: Cylindrical
Communications: Two hard-limiting repeaters (UHF and SHF)
Receiver channel band width: 50 KHz-10 MHz
Transmitter frequencies: 2200-2300 MHz

Designed and built by Hughes Aircraft Company under the direction of the US Air Force Space and Missile Systems Organization, the Tacsat spacecraft were the largest and most powerful communications satellites ever launched by the United States. Placed in geosynchronous orbit, the first of the huge two-story-high satellites was used experimentally by the US Army, Navy and Air Force for tactical communications between ground stations and mobile field units, aircraft and ships.

Above: **The two-story Tacsat was placed in synchronous orbit over the Pacific for DOD.**

Below: **The TDRS has two steerable dish antennae about 16 feet in diameter.**

TDRS

Launch vehicle: Space shuttle/IUS
First launch: 4 April 1983 (TDRS-A)
Total weight: 4700 pounds

The TDRS (Tracking and Data Relay Satellite) program spacecraft were developed by TRW in cooperation with NASA's Goddard Space Flight Center and the US Air Force Space Command. Following problems with the launch of TDRS-A in 1983, the second part of the three-satellite program, TDRS-B, was delayed. It was delivered to NASA by TRW on 8 December 1984.

TDRS-A was not placed in proper orbit due to IUS malfunction, but a series of thruster firings moved the spacecraft into proper orbit on 29 June 1983.

Teal Amber

Developed by DARPA and managed by the Air Force, Teal Amber is a military

sensor system that supersedes the Ground-based Electro-Optical Deep-Space Surveillance (GEODSS) system as a means of detecting and tracking spacecraft. A ground-base system, it is used in conjunction with spacecraft sensors systems.
(See also Teal Ruby)

Teal Ruby

The code name Teal Ruby applies to an 11-foot-long barrel-shaped telescope system utilizing a staring mosaic focal-plane array sensor system housed in an Air Force Project 888 spacecraft, which serves as the 'bus,' or carrier satellite, for the system.

Teal Ruby is an infrared sensor system designed as a space platform to use mosaic focal-plane arrays to detect aircraft in-flight over the earth's surface. Like Teal Amber, this sensor system is managed by DARPA and the US Air Force, and Rockwell International's Satellite Systems Division is the primary contractor. Originally scheduled to be carried into orbit aboard a space shuttle flight originating at Vandenberg AFB in November 1983, the first Teal Ruby deployment was postponed until 1985.

Telstar

The original Telstar, launched in 1962, was the first satellite developed and built by a private company. Telstar 1, the world's first commercial spacecraft, was a

Above: **Hughes engineer adjusts a traveling wave tube amplifier on Telstar 3.**
Below: **Artist's concept shows the deployment sequence of the USAF Project 888 spacecraft carrying the Teal Ruby experiment.**

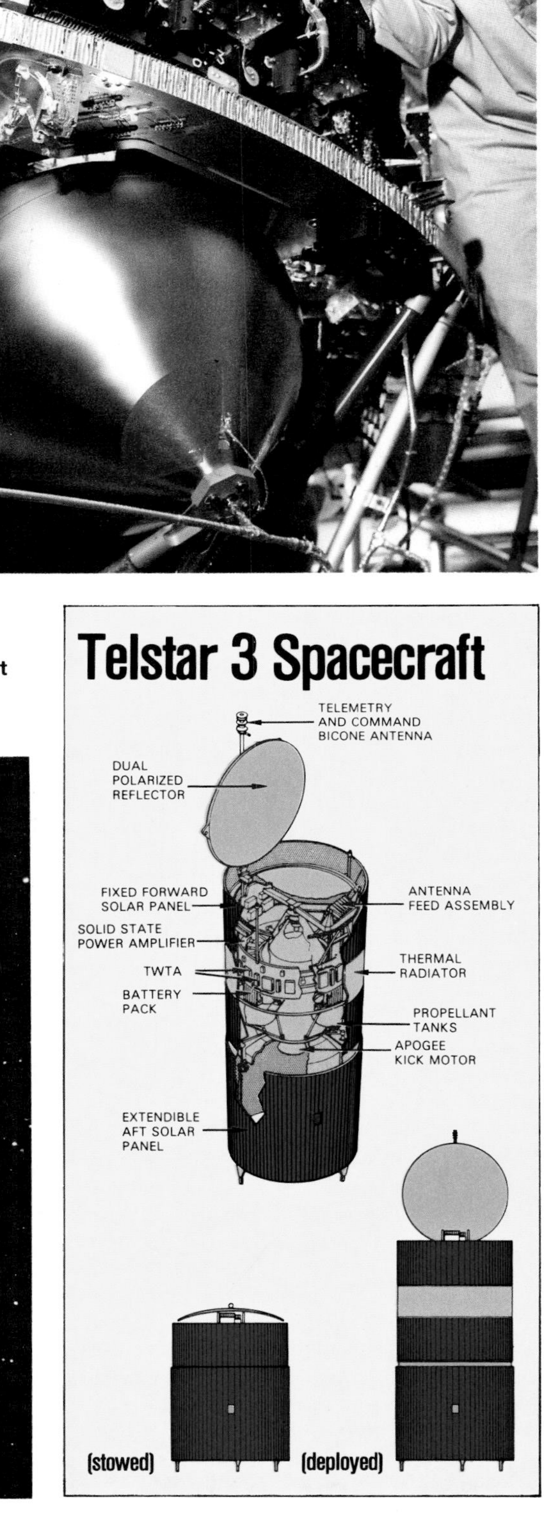

Bell's Telstar satellite served as an early microwave relay in space.

communications satellite built by the Bell Telephone Laboratories of AT&T. It is notable that a period of over 20 years separated the launches of Telstar 2 and Telstar 3. The first two Telstars were pioneering efforts in an era when man's conquest of space was new. The Telstar series came in a completely different era, by which time the use of communications spacecraft had become a routine and everyday affair. While AT&T was the owner and operator of all the Telstar spacecraft, the Telstar 3 series were built by the Space and Communcations Group at Hughes Aircraft.

Telstar 1
Launch vehicle: Thor Delta
Launch: 10 July 1962
Last transmission: 21 February 1963
Systems: Broadband active repeater receiving on 6390 megacycles and retransmitting on 4170 megacycles; 600 one-way voice channels and one TV channel; PCM/FM/AM transmitted at 136 megacycles
Shell composition: Magnesium structure and aluminum skin covered with solar cell patches, with 3600 solar cells shielded with man-made sapphire
Shape: Spherical

Telstar 2
Launch vehicle: Thor Delta
Launch: 7 May 1965
Last transmission: May 1965
Total weight: 175 pounds
Diameter: 34.5 inches
Shell composition: Aluminum
Shape: Spherical
Systems: Same as Telstar 1, plus PCM/FM/AM transmission capability at 4080 megacycles

Telstar 3
Launch vehicle: Delta (Telstar 3A)
Space shuttle (Telstar 3B, C)
Launch: 28 July 1983 (Telstar 3A)
1 September 1984 (Telstar 3B)
May 1985 (Telstar 3C)
Total weight: 7000 pounds at launch
1438 pounds in orbit
Diameter: 86 inches
Height: 108 inches (269 inches when fully deployed)
Systems: Communications and systems with 24 responder channels operating at 6/4 GHz (C band) with simultaneous telephone capacity of 21,600 calls and capable of transmitting voice, video and high-speed digitized data

The Telstar designation was revived by AT&T after two decades and assigned to a series of Hughes HS 376 spacecraft purchased by AT&T to replace satellites leased from COMSAT General Corporation. They are used to provide domestic communications services for the 48 contiguous United States, Alaska, Hawaii and Puerto Rico.

Test and Training Satellites

(See Pioneer)

Timation

Launch vehicle: Scout
Launch: May 1967 (Timation 1, four spacecraft)
30 September 1969 (Timation 2)
14 December 1971 (Undesignated, six spacecraft)
September 1972 (Triad)
14 July 1974 (Timation 3)
Total weight: 644.4 pounds (Timation 3)

A series of US Navy navigational spacecraft, the Timation (Time Navigation) program followed the Transit program and preceded the NAVSTAR Global Positioning System (GPS). Along with the Triad navigational satellites, Timation was integrated into the GPS, with Timation 3 being redesignated as Navigation Technology Satellite (NTS-1). Multiple spacecraft were placed in orbit during at least two Timation launches.
(See also NAVSTAR, Transit)

TIROS

The TIROS (Television Infrared Observation Satellite) spacecraft were the cornerstone in the NASA/ESSA/NOAA effort to develop a 24-hour television infrared weather-monitoring system. The first 10 TIROS spacecraft were followed by the TOS (TIROS Operational System) spacecraft, which were redesignated

TIROS Spacecraft

Spacecraft	Launch Date	Duration (days)	Weight (lb)	Systems
TIROS 1	1 April 1960	89	270	TV
TIROS 2	23 Nov 1960	376	280	TV
TIROS 3	12 July 1961	230	285	TV, IR
TIROS 4	8 Feb 1962	161	285	TV, IR
TIROS 5	19 June 1962	321	285	TV, IR
TIROS 6	18 Sept 1962	389	281	TV
TIROS 7	19 June 1963	1809	297	TV, IR
TIROS 8	21 Dec 1963	1287	265	APT
TIROS 9	22 Jan 1965	1238	305	TV
TIROS 10	2 July 1965	730	280	TV
TIROS-N	13 Oct 1978	(continuing)	1594	AVHRR

(See also ESSA, ITOS, NOAA)

ESSA. The TOS/ESSA spacecraft were in turn followed by a third series, originally designated ITOS (Improved TIROS Operational System) and redesignated NOAA when the National Oceanic and Atmospheric Administration took over management of the program from ESSA. The overall program finally came full circle in 1978, when ITOS/NOAA was succeeded by the TIROS-N series.

RCA was the prime contractor for the TIROS spacecraft and the project was managed by NASA. The meteorological data analysis and distribution was handled by the US Weather Bureau. All of the TIROS satellites were successfully launched and operated, except for TIROS 5, whose infrared system failed. TIROS 1 returned the first global cloud cover pictures, and subsequent spacecraft expanded on its auspicious beginning. (See also ESSA, ITOS, NOAA)

TIROS 1
Launch vehicle: Thor
Launch: 1 April 1960
Total weight: 270 pounds
Diameter: 42 inches
Height: 19 inches
Shell composition: Aluminum/stainless steel covered with solar cells
Shape: Cylindrical
Systems: Low-resolution TV camera (f 1.5 lens) and high-resolution TV camera (f 1.8 lens)

TIROS 2
Launch vehicle: Thor Delta
Launch: 23 November 1960
Total weight: 280 pounds
Specifications and description: Same as TIROS 1
Systems: Same as TIROS 1

Left: TIROS 1 transmitted 22,950 photographs.

TIROS 3
Launch vehicle: Thor Delta
Launch: 12 July 1961
Total weight: 285 pounds
Specifications and description: Same as TIROS 1
Systems: Two wide-angle vidicon cameras (f 1.5 lens) including a new infrared experiment

TIROS 4
Launch vehicle: Thor Delta
Launch: 8 February 1962
Total weight: 285 pounds
Specifications and description: Same as TIROS 1
Systems: Same as TIROS 3 but with a new lens system for one of the wide-angle cameras

Below: TIROS 8 in a simulated space atmosphere.

TIROS-N satellite during check-out.

TIROS 5
Launch vehicle: Thor Delta
Launch: 19 June 1962
Total weight: 285 pounds
Diameter: 42 inches
Height: 22 inches
Shell composition: Top and sides covered with 9260 solar cells
Shape: Cylindrical
Systems: Two vidicon cameras (one with wide-angle Elgeet lens, one with medium-angle Tegea lens), infrared horizon scanner (failed to operate)

TIROS 6
Launch vehicle: Thor Delta
Launch: 18 September 1962
Total weight: 281 pounds
Specifications and description: Same as TIROS 5
Systems: Same as TIROS 5, but without infrared system

TIROS 7
Launch vehicle: Thor Delta
Launch: 19 June 1963
Total weight: 297 pounds
Diameter: 42 inches
Height: 22 inches
Systems: Two vidicon cameras with wide-angle Elgeet lenses, and infrared experiments

TIROS 7 was an 18-sided polygon, covered with over 9000 solar cells on the top and sides, one 18-inch receiving antenna on top, and four 22-inch transmitting antennae extending from the baseplate.

TIROS 8
Launch vehicle: Delta (DSV-3B)
Launch: 21 December 1963

TIROS 9
Launch vehicle: Delta (DSV-3B)
Launch: 22 January 1965
Total weight: 305 pounds
Specifications and description: Same as TIROS 7
Systems: Two vidicon cameras

TIROS 9 was the first 'cartwheel'-configured TIROS to provide increased cloud coverage and to test the upcoming TOS system.

TIROS 10
Launch vehicle: Delta (DSV-3C)
Launch: 2 July 1965
Total weight: 280 pounds
Specifications and description: Same as TIROS 7
Systems: Same as TIROS 6

TIROS-M
(See ITOS 1)

TIROS-N
Launch vehicle: Atlas F
Launch: 13 October 1978
Total weight: 3097 pounds at launch
1594 pounds in orbit
Diameter: 74 inches
Height: 146 inches
Shape: Cylindrical, with large solar array at one end

TIROS-N carried an AVHRR to provide timely data on day and night sea-surface temperatures and ice, snow and cloud conditions.

TOPO

The TOPO project was an experimental US Army topographic triangulation navigation program. The program was managed by the Topographic Laboratory of the US Army Engineers. The TOPO spacecraft was a Secor-type spacecraft and was launched from Vandenberg AFB piggybacked with NASA's Nimbus 4 spacecraft.
(See also Secor)

TOPO 1
Launch vehicle: Thor AD/Agena D
Launch: 8 April 1970
Total weight: 48 pounds

The rectangular TOPO 1 measured 9 by 11 by 13 inches and carried a transponder and antenna system for space-ground triangulation exercises.

TOS

(See ESSA, ITOS, TIROS)

Transit

Launch vehicle: Scout
First experimental launch: September 1959 (Transit 1A)
First successful experimental launch: 13 April 1960 (Transit 1B)
First launch (operational spacecraft): December 1962 (Transit 5A)
Later Transit launches: See table
Total weight: 135 pounds (Transit 5A)
Diameter: 18 inches (without solar panels deployed)
Height: 12 inches (without antenna deployed)
Shell composition: Fiberglass
Shape: Octagonal cylinder with four solar panels extended windmill-like from cylinder
Systems: Two ultrastable oscillators operating on 150 and 400 MHz plus a digital clock

The Transit spacecraft were the first US military navigational satellites. They were originally developed by the US Navy to help ballistic-missile submarines fix their locations to an accuracy of 1.1 miles. The Transit pulse-doppler navigational spacecraft, designed for the Navy by Johns Hopkins University beginning in 1958, were launched from Vandenberg AFB and were controlled by the Naval Astronautics Group at Point Mugu NAS in California. The Transit satellites were designed to have a useful lifespan of three months, and by 1968 three series of them had been put into service in 23 successful launches. The program was declassified and made available for civilian use in 1967. At about the same time, the Transit-type spacecraft were given numerical designations in the Navsat series, which ultimately included the Transit Improvement Program (TIP) and NOVA spacecraft.
(See also Navsat, NAVSTAR, NOVA, Timation)

Later Transit Series Launches

Spacecraft	Launch date	Weight (lb)
Transit/Navsat 18	March 1968	n/a
Transit/Navsat 19	August 1970	n/a
Transit INS-1	2 September 1972	206
Transit/Navsat 20	30 October 1973	209
Transit/TIP 2	12 October 1975	352
Transit/TIP 3	1 September 1976	365
Transit/Navsat 21	28 October 1977	205
NOVA 1	15 May 1981	368

Triad
(See Timation)

US Air Force Spacecraft

(See AFSATCOM, Air Launched Miniature Vehicle, ASAT, Big Bird, Calsphere, CRL, Discoverer, DMSP, DSCS, DSPS, ELINT, Ferret, FLTSATCOM, IMEWS, Key Hole, MIDAS, NATO, NAVSTAR, Project 78, Rhyolite, SAMOS, SCATHA, Space Shuttle, Squanto Terror, Tacsat, TDRS, Teal Amber, Teal Ruby, Vela, West Ford)

US Air Force Spacecraft Project Codes

The spacecraft projects of the US Air Force are given project code numbers. Like the aircraft designations, these code numbers are assigned to all serious projects, including those in which the actual aircraft or spacecraft may not actually be built or deployed. The code designations of some of the major Air Force projects are listed below, along with the familiar names of the spacecraft.

78 Project 78 Satellite
437 Squanto Terror
467 Big Bird Satellites
647 Defense Support Program Satellites
711 Ferret Satellites
888 Teal Ruby Spacecraft 'bus'
1010 Key Hole Satellites

US Army Spacecraft

(See Courier, Explorer, Mercury (MS-1), NAVSTAR, SCORE, Secor, Space Shuttle, Tacsat, TOPO)

US Navy Spacecraft

(See FLTSATCOM, Leasat, Marisat, NAVSTAR, NOVA, Space Shuttle, Tacsat, Timation, Transit, Vanguard, Whitecloud)

Vanguard

Begun by the US Navy in 1955 in an era of bitter interservice rivalry, the Vanguard program was an attempt to place the first artificial earth-orbit satellite in space. As it turned out, the Soviet Union's Sputnik 1 and the US Army's Explorer 1 both preceded Vanguard 1 into orbit. The program included both a Vanguard launch vehicle and a Vanguard spacecraft. Like the Army-initiated Explorer satellites of the same era, the Vanguard spacecraft were generally dissimilar scientific spacecraft. They were all spherical, but their instrumentation varied, as did their size. The first three Vanguards were only 6 inches in diameter, while most of the later Vanguards were 20 inches. The ninth consisted of two spheres, 13 and 30 inches in diameter.

The Vanguard launch vehicle was based on a Viking first stage, Aerobee second stage and a specially designed third stage. The launch vehicles were given the designations TV (test vehicle) and SLV (satellite launch vehicle) to distinguish the test and operational vehicles. The first two test vehicle launches, which did not carry spacecraft, were successful, but of the 11 that followed (all carrying spacecraft) only three succeeded in placing a spacecraft into orbit. This 27 percent success rate gave the Vanguard the worst record of any US launch vehicle.

The first two Vanguard test vehicle launches (TV-1 and TV-2) took place on 1 May 1957 and 23 October 1957. These successful launches were followed by two test vehicle launches (TV-3 and TV-3 backup) carrying spacecraft. In both of

A successful Vanguard launch.

these tests, however, the launch vehicles went off course and the spacecraft did not go into orbit. Had these launches been successful, the spacecraft would have been designated Vanguard 1 and Vanguard 2. The third spacecraft to be launched *was* successful and it was designated Vanguard 1. In the Vanguard program only the three successfully orbited spacecraft were given designations (see below).

The first Vanguard launches and the first seven Vanguard spacecraft (including Vanguard 1) were under the management of the Naval Research Laboratories, after which the responsibility for the project was transferred to NASA.

Vanguard (first)
Launch vehicle: Vanguard TV-3
Launch: 6 December 1957
Total weight: 3.25 pounds
Diameter: 6 inches
Shell composition: Aluminum
Shape: Spherical
Systems: Two radio transmitters

The first Vanguard launch vehicle exploded on the launch pad.

Vanguard (second)
Launch vehicle: Vanguard TV-3 backup
Launch: 5 February 1958
Total weight: 3.25 pounds
Diameter: 6.4 inches
Shell composition: Aluminum
Shape: Spherical
Systems: Two radio transmitters

The second Vanguard launch vehicle went off course after 57 seconds and crashed without going into orbit.

Vanguard 1 (third)
Launch vehicle: Vanguard TV-4
Launch: 17 March 1958
Total weight: 3.25 pounds
Diameter: 6.4 inches
Shell composition: Aluminum, coated with a thin film of silicon monoxide
Shape: Spherical
Systems: Two radio transmitters on the inside and outside of the sphere functioned as temperature gauges

The spacecraft was successfully launched into orbit, and it continued to transmit for seven years.

Vanguard (fourth)
Launch vehicle: Vanguard TV-5
Launch: 28 April 1958
Total weight: 21.5 pounds
Diameter: 20 inches

Vanguard 1 transmitted temperature and geodetic measurements until March 1964.

Shell composition: Magnesium coated with silicon monoxide
Shape: Spherical
Systems: Radio transmitter

The fourth Vanguard experienced launch vehicle failure.

Vanguard (fifth)
Launch vehicle: Vanguard SLV-1
Launch: 27 May 1958
Total weight: 21.5 pounds
Diameter: 20 inches
Shell composition: Magnesium coated with silicon monoxide
Shape: Spherical
Systems: Two radio transmitters

The fifth Vanguard also experienced launch vehicle failure.

Vanguard (sixth)
Launch vehicle: Vanguard SLV-2
Launch: 26 June 1958
Total weight: 21.5 pounds
Diameter: 20 inches
Shell composition: Magnesium coated with silicon monoxide
Shape: Spherical
Systems: Radio transmitter

The sixth Vanguard suffered launch vehicle failure.

Vanguard (seventh)
Launch vehicle: Vanguard SLV-3
Launch: 26 September 1958
Total weight: 21.5 pounds
Diameter: 20 inches
Shell composition: Magnesium
Shape: Spherical
Systems: Two radio transmitters and infrared-sensitive photo cells to scan earth cloud cover

Launch vehicle second-stage failure of the seventh Vanguard resulted in re-entry after only one orbit.

Vanguard 2 (eighth)
Launch vehicle: Vanguard SLV-4
Launch: 17 February 1959
Total weight: 20.7 pounds
Diameter: 20 inches
Shell composition: Magnesium, coated with highly polished silicon monoxide
Shape: Spherical
Systems: Two photocell units, recorder, radio transmitter and receiver

Vanguard 2 was the second successful Vanguard launch and the first Vanguard managed by NASA. It returned the first photo of earth from a satellite, and continued to return data for 27 days.

Vanguard 3 (ninth)
(Provisionally designated Vanguard 3, but this was canceled and reassigned)
Launch vehicle: Vanguard SLV-5
Launch: 13 April 1959
Total weight: 23.3 pounds
Shell composition: Vanguard 3A: fiberglass and phenolic resin
Vanguard 3B: laminated aluminum foil on a plastic sheet

Two independent spheres 13 and 30 inches in diameter were designated Vanguard 3A and 3B, respectively. They were connected by a cylinder 17.5 inches long and 2.5 inches wide. Vanguard 3B, an inflatable sphere for optical tracking, carried no systems, while Vanguard 3A carried a precise magnetometer to map the earth's magnetic field.

The ninth Vanguard experienced launch vehicle failure.

Vanguard (tenth)
Launch vehicle: Vanguard SLV-6
Launch: 22 June 1959
Total weight: 21.5 pounds
Diameter: 20 inches
Shell composition: Magnesium alloy
Shape: Spherical
Systems: Two radio transmitters, solar temperature gauge

The tenth Vanguard suffered launch vehicle failure.

Vanguard 3 (eleventh)
Launch vehicle: Vanguard SLV-7
Launch: 18 September 1959
Total weight: 50.7 pounds
Diameter: 20 inches
Shell composition: Magnesium
Shape: Spherical, with magnetometer
Systems: Proton-precession magnetometer, micrometeoroid detectors, two solar X-ray ionization chambers, radio transmitters

Vanguard 3 was the third successful Vanguard launch and the last attempted launch of the Vanguard series. It surveyed the lower edge of Van Allen Belt and provided a comprehensive survey of the earth's magnetic field, transmitting until 8 December 1959.

Vela

The objective of the US Air Force Vela spacecraft program was to detect nuclear tests in space. Although planned in the 1950s, the purpose and first launch coincided with the signing of the 1963 Nuclear Test Ban Treaty. The treaty specifically banned the testing of nuclear weapons in space and Vela could be used to monitor compliance with such treaties.

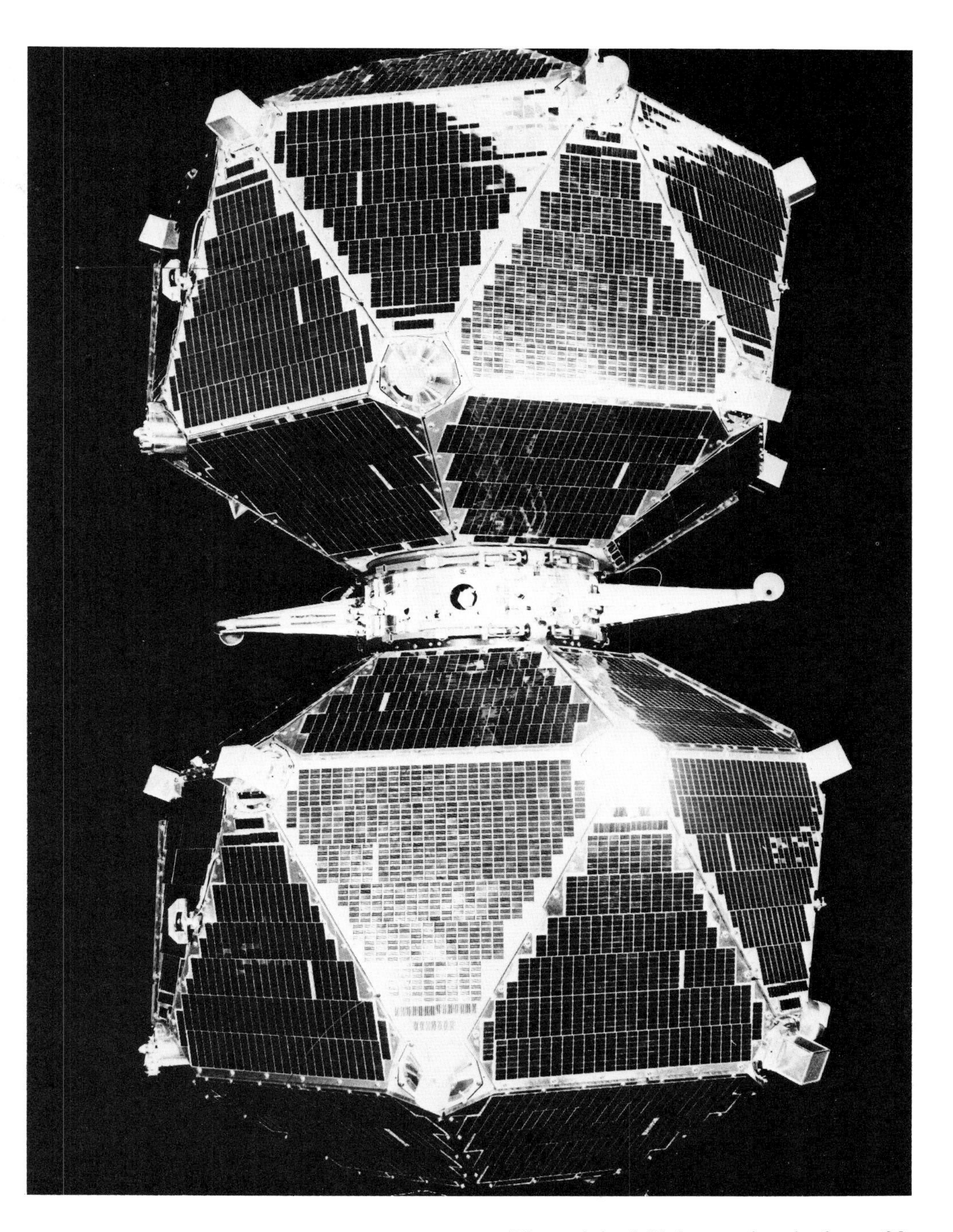

Vela satellites in twin-launch configuration.

The spacecraft's sensors can detect nuclear explosions on earth, in near-earth space and as far out as Mars or Venus. The spacecraft were developed by TRW and designed to be launched in pairs. The final pair, Vela 11 and Vela 12, continued to operate well into the 1980s.

Vela (Original developmental block)
Launch vehicle: Atlas Agena D
Launch: 17 October 1963 (Vela 1 and 2)
17 July 1964 (Vela 3 and 4)
20 July 1965 (Vela 5 and 6)
Total weight: 508.2 pounds
Diameter: 55.38 inches

The original Vela consisted of two 20-sided polyhedrons connected by a central cylinder containing an apogee motor.

Vela (Upgraded block)
Launch vehicle: Titan 3C
Launch: 28 April 1967 (Vela 7 and 8)
23 May 1969 (Vela 9 and 10)
Weight and description: Similar to earlier Velas

Vela (Advanced operational block)
Launch vehicle: Titan 3C
Launch: 8 April 1970 (Vela 11 and 12)
Total weight: 571 pounds
Diameter: 49.53 inches

The advanced Velas consisted of two 26-sided polyhedrons connected by a central cylinder containing an apogee motor.

Viking

The objective of the Viking program was to insert two spacecraft into orbit around the planet Mars, and then release from each a landing vehicle that would conduct a soft landing on the surface of the red planet. The project began on 15 November 1968 and culminated in successful soft landings on the Martian surface by both Viking landers in July and September of 1976. Two Viking orbiters and two Viking landers were employed because the chances of success were such that a backup spacecraft was considered prudent. Both spacecraft were very successful, however, and made an extremely thorough and extensive survey of Mars, mapping 97 percent of its surface.

While they were not the first spacecraft to attempt a soft landing on another planet, the Viking landers (VL) were by far the most successful. In several attempts, no Soviet or American soft-lander had up to then survived more than 65 minutes on Venus. In several attempts, the Russians had yet to put a successful soft-lander on Mars, although their Mars 3 had survived for 20 seconds on the surface in 1971.

Mars has always been of great interest to scientists because of its similarity to earth and because it seems to be the most likely other planet to be able to support life. Thus, the search for Martian life became an important part of the Viking project, and three biology experiments were designed to attempt that task. To prevent contamination of the Martian surface with earth organisms, the landers were sterilized and encapsulated in a bioshield, which was discarded in the vacuum of space.

The Viking exploration of Mars turned up no artifacts of a Martian civilization and detected no sign of the canals that were once thought to exist there. It did discover that the atmosphere had once been thicker, the surface warmer and that there had once been large rivers (containing water) on the planet. No evidence of past or present life was detected, but chemical analyses did open the door to that possibility. The atmosphere of Mars, while only 1.0 percent as thick as that of earth, was discovered to contain those elements (nitrogen, carbon, oxygen and water vapor) that sustain life on earth. A great deal of water vapor was found in the northern hemisphere, with considerable water frozen in the north polar ice cap. Morning ground fog was seen around the landers, and subsurface permafrost was detected.

Viking 1
Launch vehicle: Titan 3/Centaur
Launch: 20 August 1975
Insertion into Martian orbit: 19 June 1976
Soft landing on Mars: 20 July 1976
Last transmission, orbiter: 7 August 1980
Last transmission, lander: November 1982

On 14 July 1976, Viking 1 was maneuvered into synchronous orbit over Chryse Planitia, a landing area that its own cameras had helped mission control to locate. Four days later, the VL-1 lander

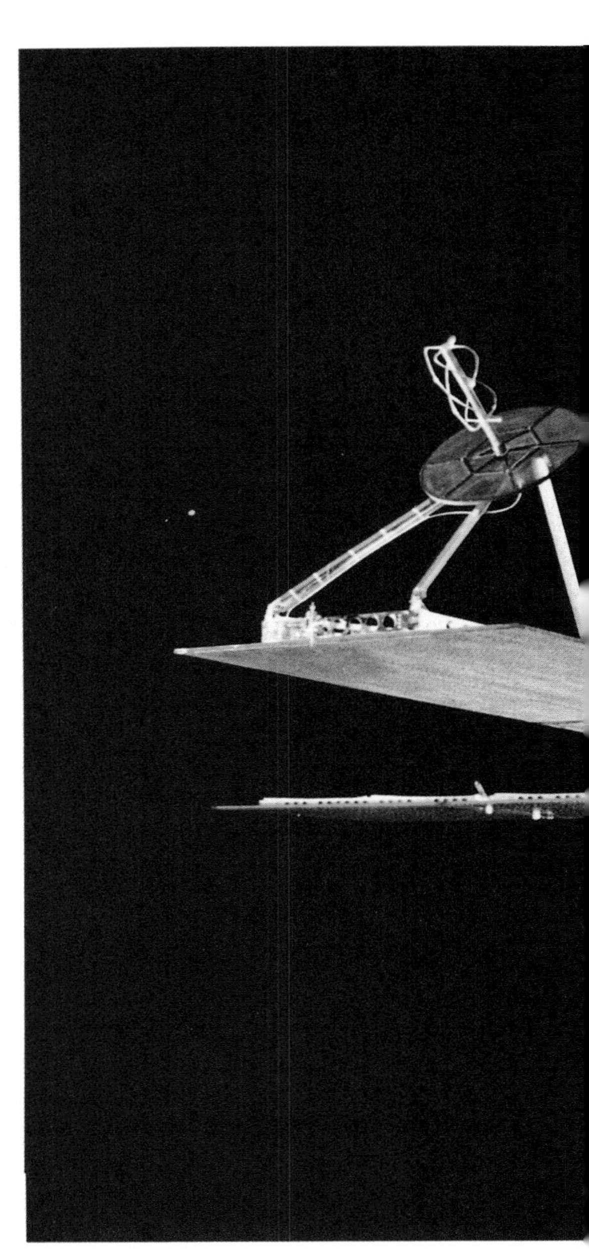

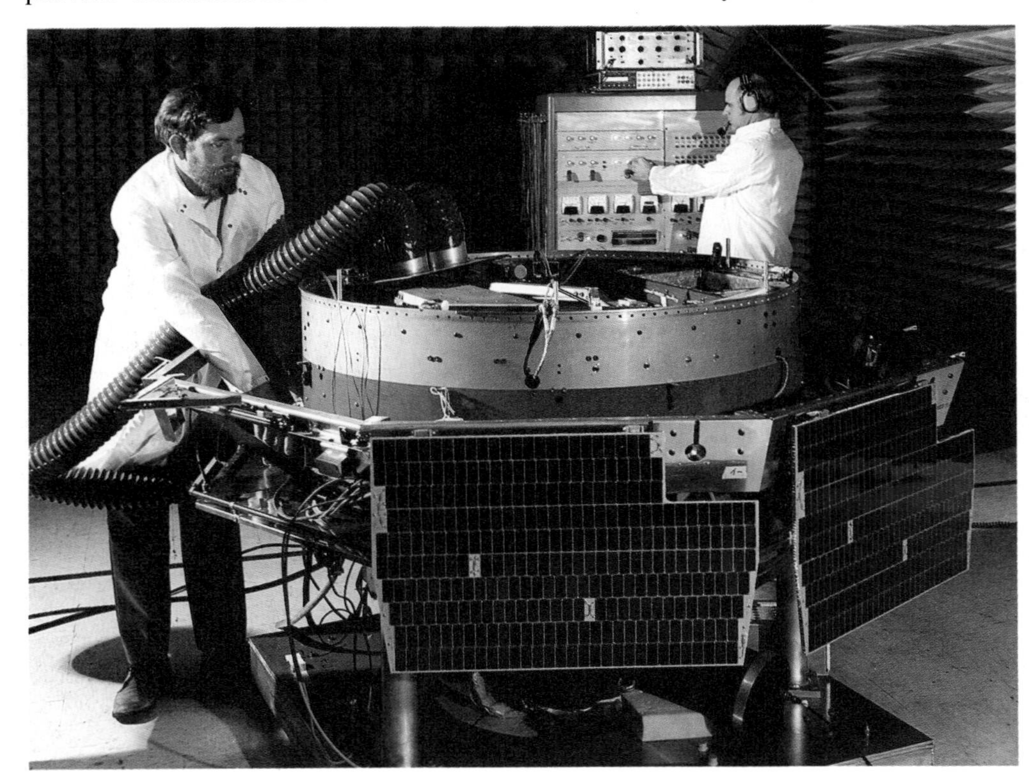

Top: Viking Mars orbiter and lander capsule.
Right: Viking lander capsule No 1 at KSC is de-encapsulated and propellant tanks are installed (*above*) before decontamination.
Left: Viking satellite platform's electrical systems are tested in an anechoic chamber.

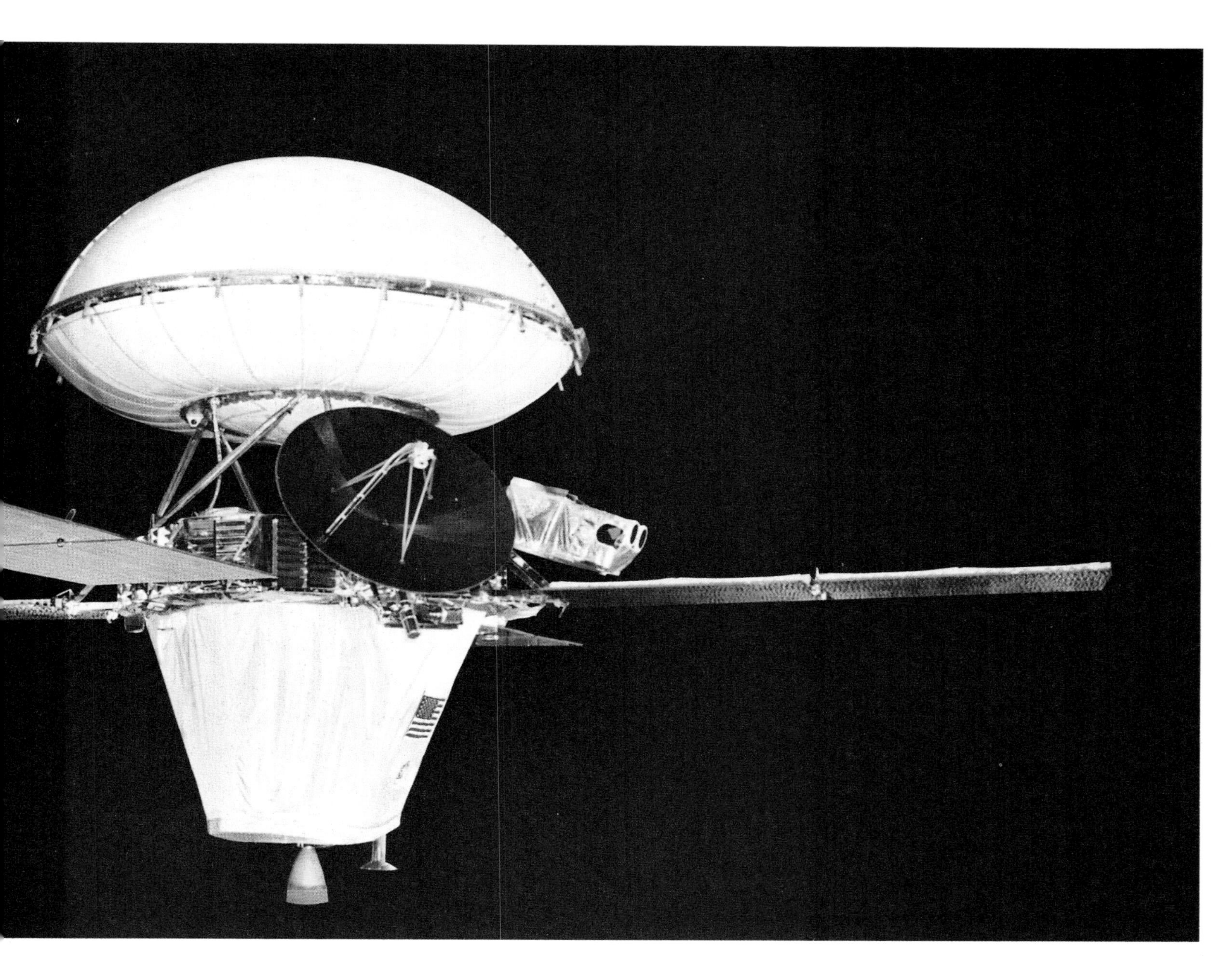

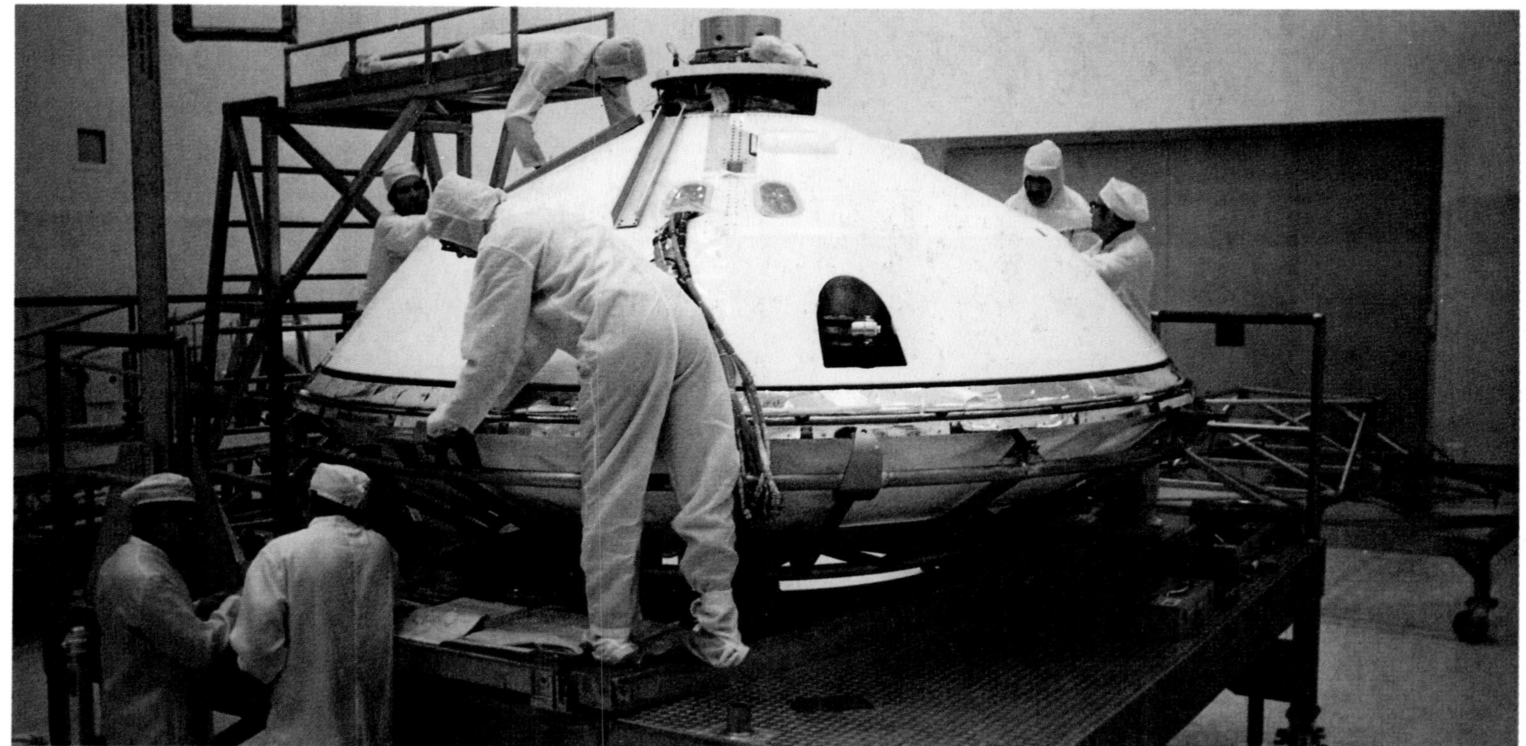

Viking Spacecraft, Systems

Systems	Function
Orbiter:	
Two vidicon cameras	(Orbiter imaging)
Infrared spectrometer	(Water vapor mapping)
Infrared radiometer	(Thermal mapping)
Lander:	
Entry science	
Retarding potential analyzer	(Ionospheric properties)
Mass spectrometer	(Atmospheric composition)
Pressure, temperature and acceleration sensors	(Atmospheric structure)
Radio science	
Orbiter and lander radio and radar systems	(Celestial mechanics, atmospheric properties, test of general relativity)
Two facsimile cameras	(Lander imaging)
Biological analyses	
Three separate experiments, gas exchange, labeled release and pyrolytic release were included to test different biologic models	(Metabolism, growth, photosynthesis)
Molecular analysis	
Gas chromatograph mass spectrometer	(Organic compounds)
Mass spectrometer	(Atmospheric composition)
X-ray fluorescence spectrometer	(Inorganic analysis)
Pressure, temperature and wind velocity sensors	(Meteorology)
Three-axis seismometer*	(Seismology)
Magnets on sampler and a camera test chart, observed by cameras	(Magnetic properties)
Various engineering sensors	(Physical properties)

*Operable on Viking Lander 2 only

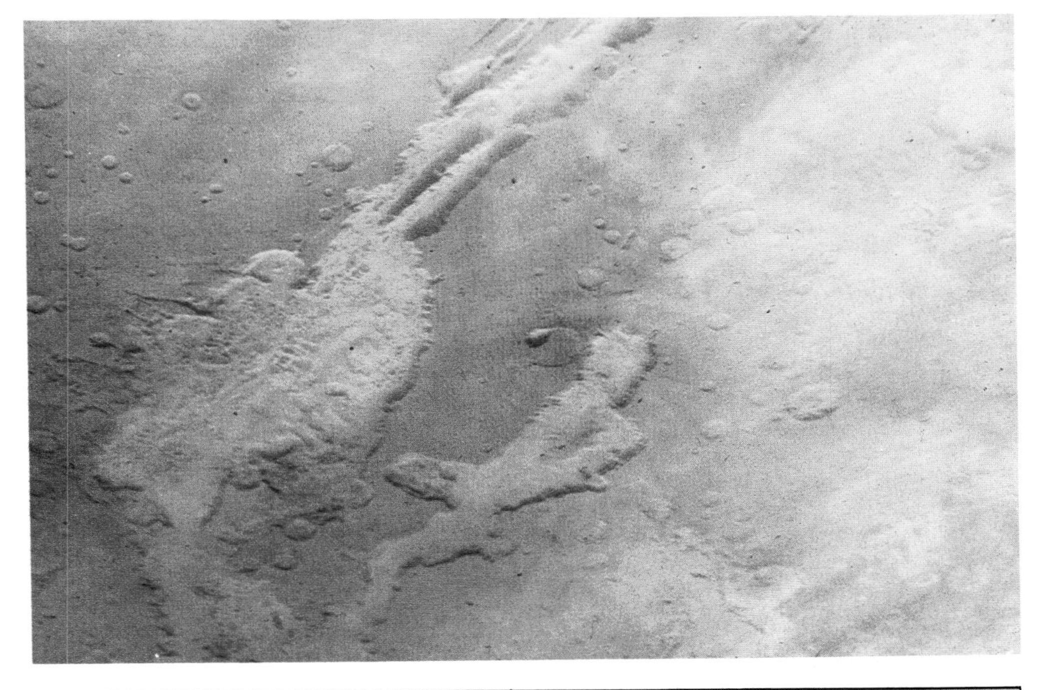

Above: **The Viking 1 lander took this image of Mars and discovered various geologic features.**
Left: **Viking lander capsule No 1 is prepared for installation of flight subsystems.**
Above right: **Mars in late winter and Valles Marineris (*right*) from 19,200 miles' altitude.**

was separated from the orbiter and it began its descent to the Martian surface. A combination of terminal descent rockets and a 52-foot-diameter parachute slowed VL-1 for its successful landing. Transmission of photos began 25 seconds after the landing and the pictures reached earth 19 minutes later. The pictures revealed a reddish rocky landscape with no sign of life. The sampler arm was deployed on 25 July and on 28 July the first Martian soil sample was pulled into VL-1's biology experiment area for analyses. These experiments attempted to stimulate and detect any life forms that might exist in the soil. An unpredictably high level of oxygen was released when the samples were stimulated by sunlight or water, but not if the sample was sterilized. One possible reason for this release of oxygen was that plants were present and undergoing photosynthesis. Another possible answer was that a purely inorganic chemical reaction was taking place. Neither of these possibilities could be proven or disproven conclusively, leaving open the question of life on Mars.

The Viking 1 orbiter continued to return data about Mars, and in May 1977 it encountered the Martian moon, Phobos. In November 1980, the VL-1 lander was renamed Mutch Memorial Station, after NASA's Dr Thomas Mutch, former Viking lander imaging team leader who had disappeared in the Himalayas.

Viking 2
Launch vehicle: Titan 3/Centaur
Launch: 9 September 1975
Insertion into Martian orbit: 7 August 1976
Soft landing on Mars: 3 September 1976
Last transmission, orbiter: 25 July 1978
Last transmission, lander: 12 April 1980

Viking Spacecraft, General Specifications

	Orbiter	**Lander**
Weight	5125 lb	2353 lb
Shape	Irregular octagonal prism with solar arrays (similar to Mariner)	Hexagonal body with three landing legs
Diameter	85 by 99.2 inches	59 inches
Height	114 inches	76 inches
Width (fully deployed)	396 inches (across solar arrays)	144 inches (across legs)
Power source (systems)	23,250 square inches of solar cells, plus two nickel-cadmium batteries	SNAP-19 radioisotope thermoelectric generators
Power source (propulsion)	One 300-lb thrust maneuvering and orbit-insertion engine	Three 585-lb thrust terminal descent engines with 18-nozzle clusters to diffuse blast on surface

As had been the case with Viking 1, the orbiter helped locate the landing area in Utopia Planitia, and the spacecraft was placed in synchronous orbit over the site on 27 August. The successful landing of VL-2 was followed by deployment of the seismometer, a system that had jammed on VL-1. There was little seismic activity detected, despite the presence of enormous volcanic mountains. The terrain of the landing site was similar to that of VL-1, though lower in elevation. The weather at the site was clearer and warmer. The first soil samples were taken on 12 September, but the results of the VL-2 biology experiments differed from those of VL-1 and showed no sign of life. VL-2 was shut down on 10 November 1976 after returning enormous quantities of data about the Martian surface, but the question of Martian life was deferred.

The Viking 2 orbiter continued in service and was used to conduct a close survey of the Martian moon, Derimos, in September 1977. The orbiter was finally powered down on 25 July 1978.

Voyager

The Voyager program developed by NASA's Jet Propulsion Laboratory centered on two identical spacecraft whose mission was to conduct a scientific investigation of Jupiter and Saturn, and to expand the information gained from Pioneer 10 and 11 about those two planets. It was also to conduct flybys of Uranus and possibly Neptune, planets so far distant from earth that no clear pictures had ever been obtained. The photos obtained by the two Voyagers of Jupiter, Saturn and their respective moons were astounding. Our understanding of these two complex celestial bodies was vastly enhanced by the close approaches of the two spacecraft, which included passing through Saturn's ring system.

The surveys conducted by the Voyagers provided a wealth of new discoveries, including answers that had eluded astronomers for centuries. Close-up pictures were returned of Jupiter's great red spot. Close-ups of its moons showed that Io was waterless, the extremely volcanically active Europa was smooth and abundant in water, and Callisto and Ganymede consisted equally of water ice and rock. Saturn's rings and most of its moons were shown to be largely water ice as well, except for Phoebe (the outmost moon), which is mostly rock, and Titan (the largest), which was about half rock.

Voyagers 1 and 2 were identical.

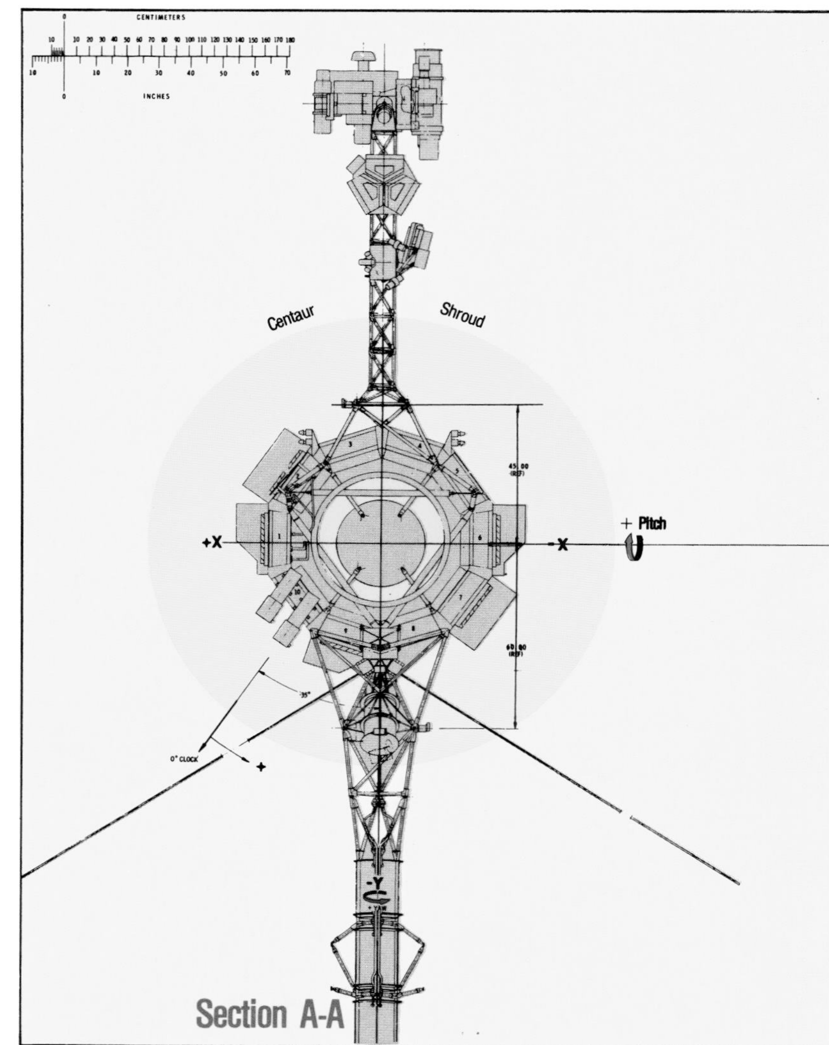

Above: **This Voyager test model was used for electrical and countdown tests. *Right:* Voyager 1 photographed Saturn and its satellites Tethys, above, and Dione.**
This photograph (*far right*) was taken of Saturn from a distance of 11 million miles.

Voyager 1 managed to pass within 2500 miles of Titan on 11 November 1980. Titan, which, like Jupiter's Ganymede, is larger than the planets Mercury and Pluto, was found to have an atmosphere of 82 percent nitrogen (78 percent of earth's atmosphere is nitrogen). While the balance of earth's atmosphere is oxygen, the rest of Titan's is composed of ethane, methane acetylene, cyanide and other organic compounds. With Titan's surface temperature a chilly −288 degrees Fahrenheit, the abundant water exists as solid ice, but methane can exist as a liquid. To quote NASA's interpretation of this data, 'On Titan then, methane rain or snow may fall from methane clouds. Methane rivers may flow through methane channels and methane oceans may fill icy basins.'

Jupiter and Saturn have turbulent atmospheres, with storms that have raged for centuries, large enough to swallow the

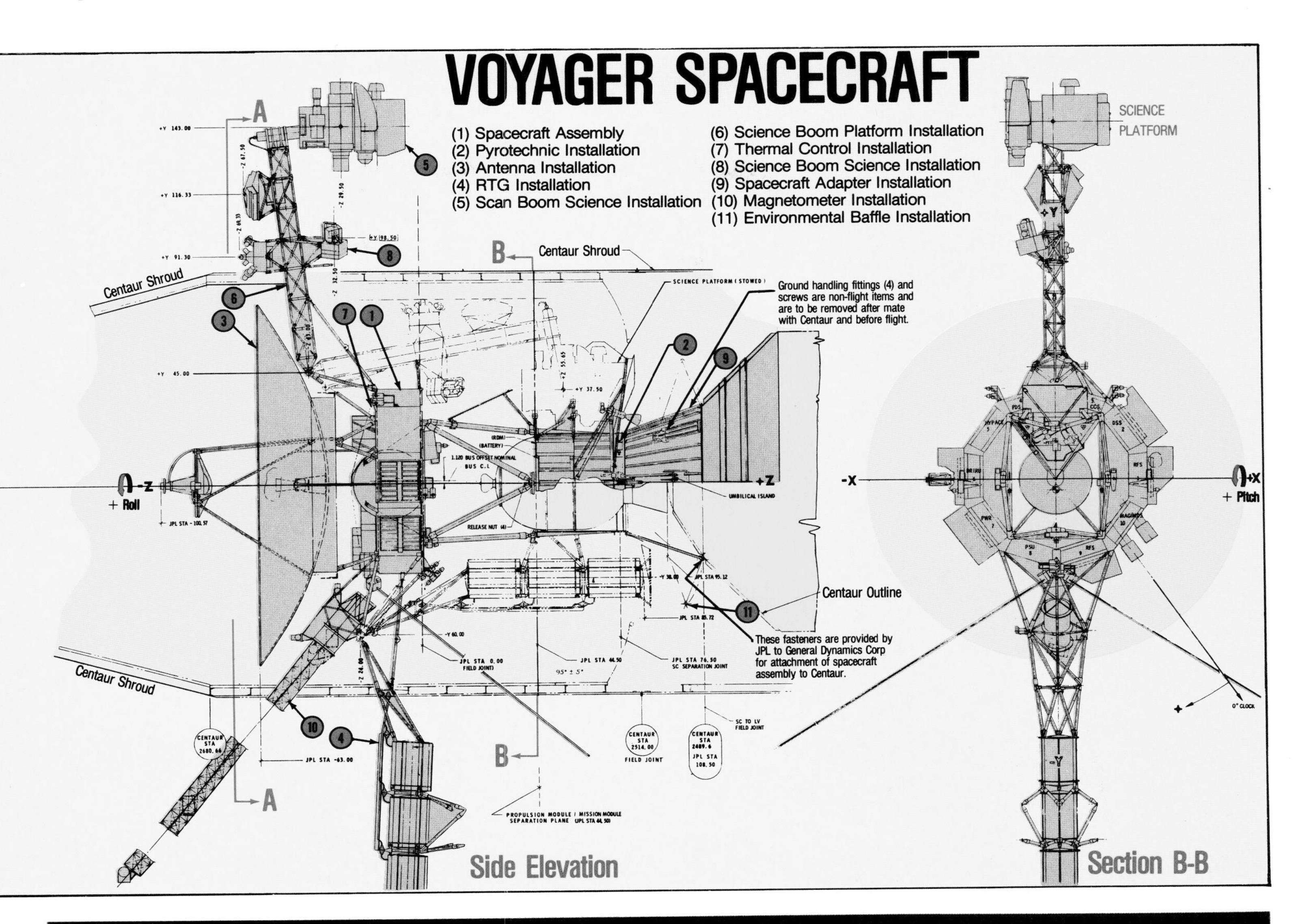
VOYAGER SPACECRAFT
(1) Spacecraft Assembly
(2) Pyrotechnic Installation
(3) Antenna Installation
(4) RTG Installation
(5) Scan Boom Science Installation
(6) Science Boom Platform Installation
(7) Thermal Control Installation
(8) Science Boom Science Installation
(9) Spacecraft Adapter Installation
(10) Magnetometer Installation
(11) Environmental Baffle Installation
Centaur Shroud
Ground handling fittings (4) and screws are non-flight items and are to be removed after mate with Centaur and before flight.
Centaur Outline
These fasteners are provided by JPL to General Dynamics Corp for attachment of spacecraft assembly to Centaur.
+ Roll
+ Pitch
SCIENCE PLATFORM
Side Elevation
Section B-B

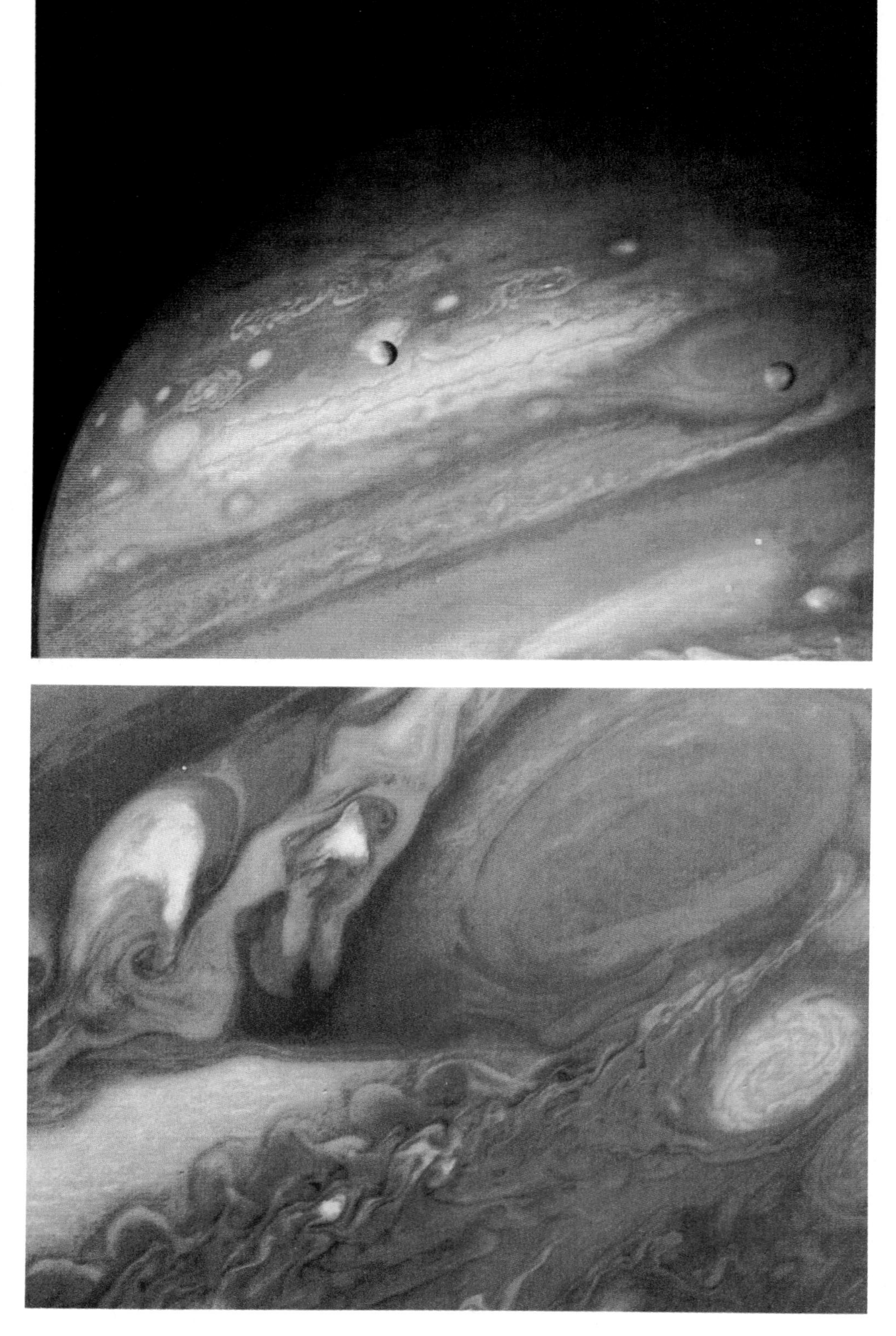

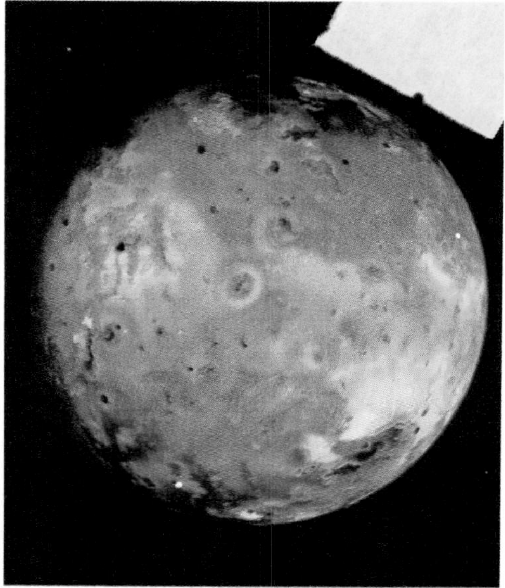

Above: **Voyager 1 neared Jupiter's satellite Io on 4 March 1979 and returned this image.**
Left: **Satellite Io, left, is composed mostly of sulfur; Europa, right, has a thin ice crust.**

Above: **Voyager 1 photographed Jupiter's turbulent atmosphere and Great Red Spot, *upper right*.**

earth. Indeed, the two planets are entirely atmosphere, with no detectable surface. Within this atmosphere, heat is chemically generated so that both planets radiate 250 percent as much heat as they receive from the distant sun.

Because the Voyagers are operating so far from the sun, solar energy is too weak to generate electricity, so the Voyager spacecraft are powered by small nuclear-power generators. This system will keep them operating beyond the encounters with Uranus and Neptune, and will permit them to transmit data to earth from interstellar space until about AD 2007, at which time they will be roughly nine billion miles from earth. Each of the Voyager spacecraft carry with them a 'sounds of earth' recording, which includes greetings in 60 languages, music and sounds of animals, the ocean and the wind.

Voyager 1
Launch vehicle: Titan 3/Centaur
Launch: 5 September 1977 (behind Voyager 2)
Closest encounter with Jupiter: 5 March 1979 (177,720 miles)
Closest encounter with Saturn: 12 November 1980 (77,000 miles)
Encounter with Uranus: None expected
Encounter with Neptune: None expected
Total weight: 46,000 pounds at launch (including 230 pounds of scientific payload)
Jupiter trajectory weight: 1797 pounds
Antenna diameter: 142.7 inches

Voyager 1 was equipped with a radio science experiment to obtain occultation and celestial mechanics data, plus 10 scientific experiments including: infrared spectroscopy and radiometry; ultraviolet spectroscopy; photopolarimetry; plasma detection; low-energy charged particle detection; magnetometry; planetary radio astronomy and plasma wave detection; and imaging (two TV cameras, one with a narrow-angle 1500 mm f8.5 lens, one with a wide-angle 200 mm f3 lens).

Voyager 2
Launch vehicle: Titan 3/Centaur
Launch: 20 August 1977
Closest encounter with Jupiter: 9 July 1979 (399,560 miles)
Closest encounter with Saturn: 25 August 1981 (63,000 miles)
Encounter with Uranus: 30 January 1986
Encounter with Neptune: September 1989
Weight and specifications: Same as Voyager 1
Systems: Same as Voyager 1

Westar

The Westar spacecraft were the first US domestic sychronous-orbit communications satellites. They are Hughes HS 376 spacecraft built for Western Union by Hughes Aircraft, and are similar to the Hughes-built Anik communications spacecraft built for Telesat of Canada. Western Union was one of the first applicants to obtain approval from the US Federal Communications Commission to operate a domestic satellite system. The Westar spacecraft serve the continental US, Alaska, Hawaii and Puerto Rico from the primary Western Union earth station at Glenwood, New Jersey and from other Westar earth stations located near Atlanta, Chicago, Dallas and Los Angeles. The 5-foot parabolic antenna was used to illuminate the contiguous 48 states. Antenna phasing provided coverage of Alaska and Puerto Rico, while a separate spot beam was used for Hawaii.

The first five Westar launches were successful. The sixth, which took place from the space shuttle in February 1984, failed to achieve geosynchronous orbit, but it was rescued by a space shuttle crew on 14 November 1984 and returned to earth for refurbishment and relaunch.

All of the Westar spacecraft were launched by NASA on a reimbursable basis.

(See also HS 376)

Westar 1
Launch vehicle: Delta 2914
Launch: 13 April 1974
Total weight: 674 pounds in orbit
Diameter: 75 inches
Height: 135 inches
Shape: Cylindrical
Transponders: 12 (each with 36 MHz bandwidth)
Frequency: 4-6 GHz

Westar 2
Launch vehicle: Delta 2914
Launch: 10 October 1974
Weight and specifications: Same as Westar 1
Systems: Same as Westar 1

Right: Westar 1 prepared for launch. *Below:* Westar 2 receives final touches and Westar 3 is checked.

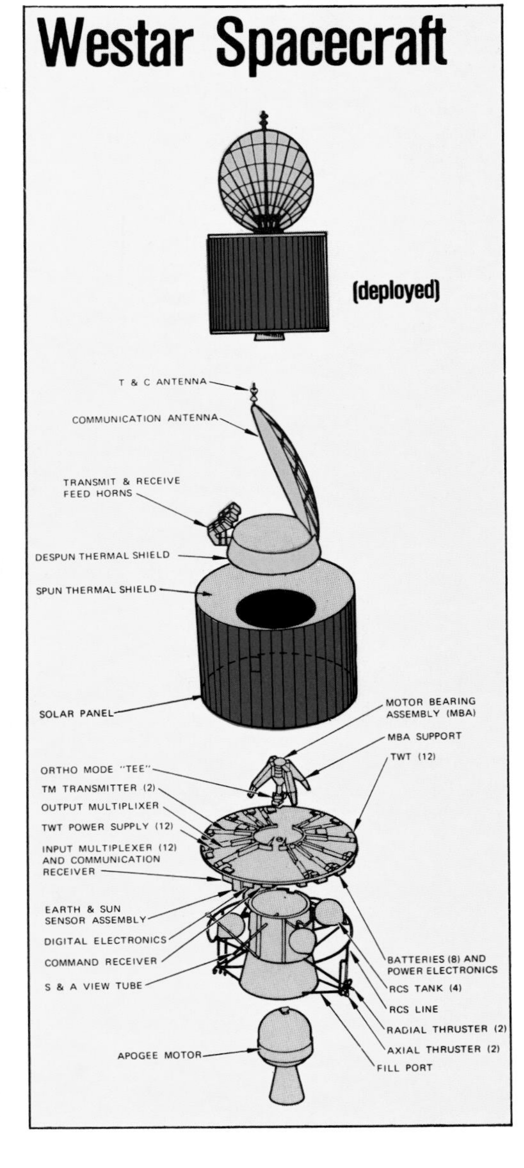

Westar 3
Launch vehicle: Delta 2914
Launch: 9 August 1979
Weight and specifications: Same as Westar 1
Systems: Same as Westar 1

Westar 4
Launch vehicle: Delta 2914
Launch: 25 February 1982
Total weight: 1441 pounds
Diameter: 85 inches
Height: 112 inches (stowed) 260 inches (deployed)
Shape: Cylindrical
Transponders: 12 (each with 36 MHz bandwidth)
Frequency: 4-6 GHz

Westar 5
Launch vehicle: Delta 2914
Launch: 8 June 1982
Weight and specifications: Same as Westar 4
Systems: Same as Westar 4

Westar 6
Launch vehicle: Space shuttle/PAM-D
Weight and specifications: Same as Westar 4
Systems: Same as Westar 4

West Ford

In 1963 the development of communications spacecraft was in its early stage. Passive spacecraft, such as Echo, were used to bounce radio signals between earth stations. The US Air Force came up with the idea of placing 500 million tiny copper wires in space to relay radio messages. Astronomers feared that the metal would inhibit radio astronomy and they opposed the idea. Nevertheless, the USAF went ahead in 1963. The results were disappointing for the Air Force because, to quote Thomas Karas, 'The copper cloud drifted apart, making for neither good communications nor bad astronomy.'

Whitecloud

Launch vehicle: Atlas
First launch: 30 April 1976 (Whitecloud 1)
(Remaining data is still classified)

The Whitecloud spacecraft are a series of supersecret ocean-surveillance satellites used to monitor the movement of Soviet ships and submarines. Developed for the Naval Research Laboratory by Martin Marietta, they are placed in 700-mile-high orbits, where they release three smaller spacecraft that fan out to cover a wider area. The spacecraft follow the movement of ships and submarines by tracking their radar and communications. The project was so secret that Naval Research Laboratory personnel were not even permitted to use the word in telephone conversations – until souvenir postal covers with Vandenberg AFB postage cancelations showed up for sale at the tourist gift shop in the cafeteria at NASA's Johnson Space Center in October 1984.

Dale Gardner approaches the spinning Westar 6 (*left*), achieves a hard dock and uses a 'stinger' device to stabilize it (*below*), Joseph Allen helps bring Westar 6 into the cargo bay (*above*).

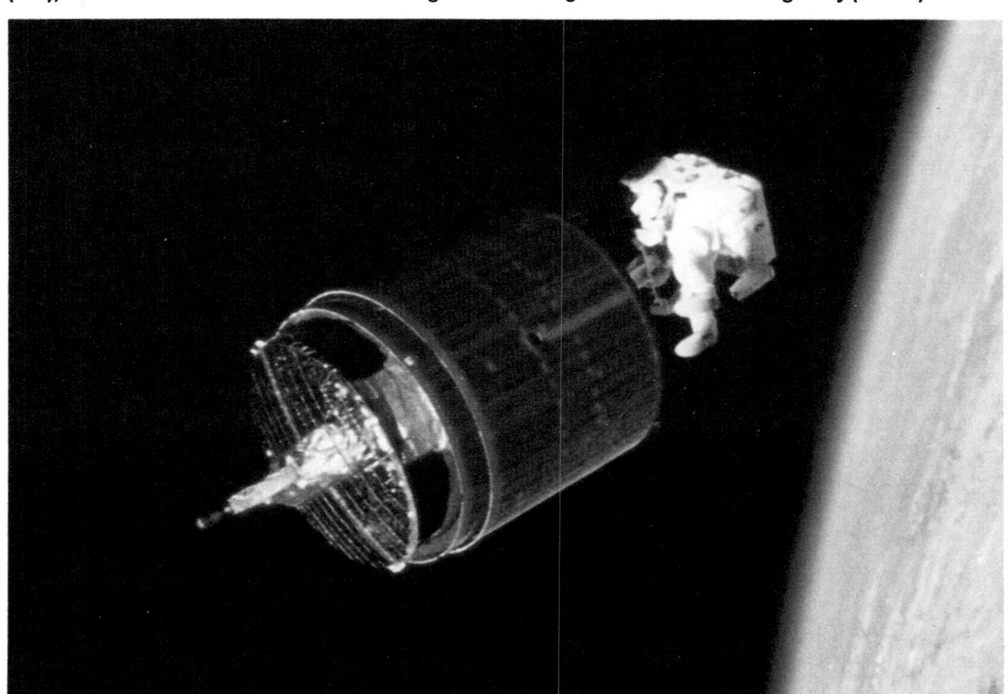

Appendix 1

Space Stations

The establishment of permanently manned space stations has been a dream of science fiction writers for years, until the early 1960s. With the advent of manned spaceflight, it became the dream of serious aerospace scientists, both in the US and the Soviet Union. The Soviets led the way with the launch of the first of their Salyut series in 1971. There were six Salyut space stations successfully placed in orbit prior to Salyut 7 in 1982. With its series of Salyuts, the Soviet Union has managed to maintain a presence in space for long blocks of time, setting many space endurance records in the process. Some individual cosmonauts have served aboard Salyut for over six months at a stretch. Valery Ryumin, for example, served two tours of duty of 175 and 185 days, or a total of nearly a year in the space station.

While many designs proliferated, the first serious US space station project was the Manned Orbital Laboratory (MOL) anounced by President Lyndon Johnson on 26 August 1965. The MOL project was developed by Douglas Aircraft for the US Air Force and was designed to interface with the Gemini manned spacecraft developed by McDonnell Aircraft. Coincidentally, the two firms merged midway through MOL's development.

The MOL was a simple cylinder 40 feet long and 10 feet in diameter that was launched into space on a Titan 3C launch vehicle, with a manned Gemini capsule on top of it. The MOL was designed to support the two-man Gemini crew in a shirtsleeves environment for up to 30 days. The Air Force went as far as to launch a full-scale MOL mock-up, complete with an unmanned Gemini capsule, on 3 November 1966.

The MOL project, however, was born in a difficult era. The important Apollo lunar program overshadowed nearly all other US space programs and made Gemini seem obsolete. In the meantime the Vietnam War put a severe strain on the Air Force budget and MOL was deployed. The project was finally canceled in August 1969.

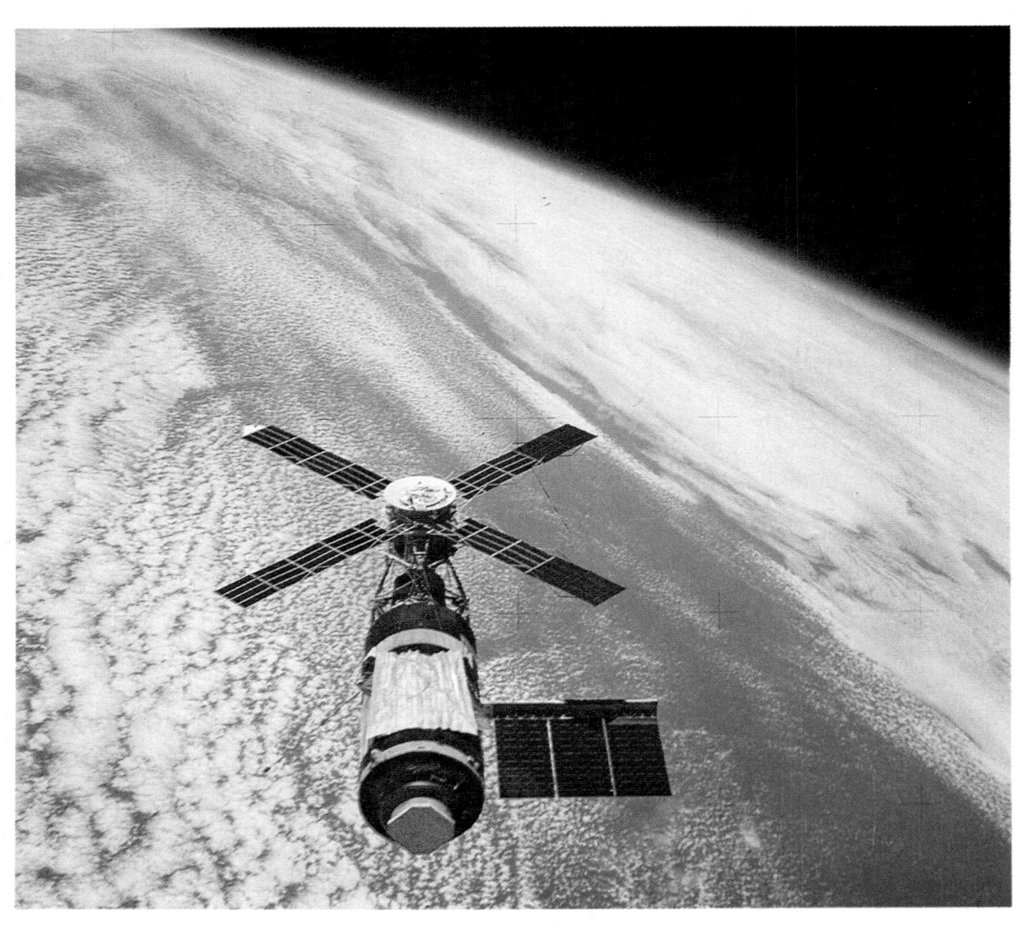

Above: The Skylab space station cluster in orbit. The left solar shield was lost and the one on the right screens the OWS.
Right: A two-stage Saturn 5 rocket launched the Skylab payload to earth orbit in May 1973.

Skylab

America's first space station, and the only one that it deployed during the first three decades of manned spaceflight, was the 100-ton Skylab. The main part of it, the orbital workshop, was fashioned from the fourth stage (S-4B) of a Saturn 5 launch vehicle of the type used for the Apollo lunar missions. It was designed to be served by Apollo spacecraft like those used on the lunar missions, and when the two were docked the Apollo CSM functioned as part of the space station. The interior volume of Skylab was 11,334 cubic feet, roughly four times that of the Soviet Salyut space station, and roughly the size of a small house.

Designed and built by McDonnell Douglas Astronautics, Skylab was placed into earth's orbit on 14 May 1973. It was a cluster of four units, three of them habitable. These included the orbital workshop (OWS) containing the principal crew quarters and work areas; the airlock module (AM) containing the station's control and monitoring center as well as access to the outside for extravehicular activity; the Apollo telescope mount (ATM), a solar observatory; and the multiple docking adapter (MDA), which contained docking ports for the Apollo spacecraft that brought astronauts to Skylab and controls for the ATM and other scientific equipment. The OWS, converted from the Saturn fourth stage, was 21.6 feet in diameter, 48.1 feet long, and divided into two stories. The upper of these was equipped with such items as

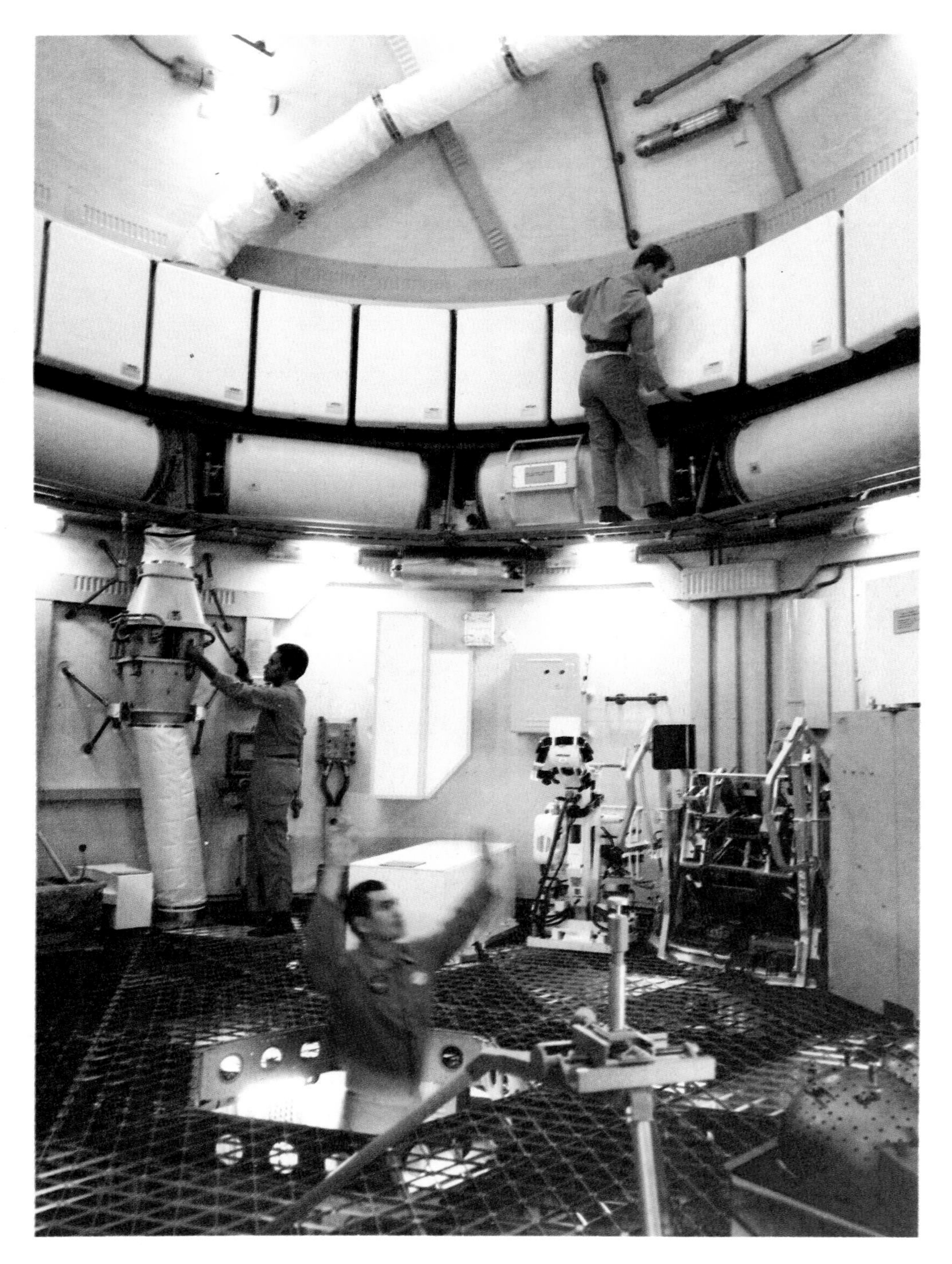

The interior of the Skylab space station during tests on earth.

food lockers, refrigerators, water tanks and space-suit lockers. The lower story contained crew quarters and an experiment station. The habitable volume of the OWS was more than 10,000 cubic feet.

The AM, with a volume of 622 cubic feet, was 17.6 feet long, 10 feet in diameter and permitted access to the exterior without the need to depressurize the entire station. The MDA, just a few inches shorter than the AM, was the control center for the ATM. It was the first orbiting solar observatory controlled by men in outer space, although ground control was exercised during crew sleep periods and when the station was not staffed.

The objectives of the Skylab program were to evaluate long-term effects of spaceflight on people and equipment, to conduct scientific observations of both the earth and sun, and to provide information for the development of future space stations. In addition to cameras and film, scientific equipment included instruments for white-light coronagraphy, an ultraviolet scanning polychromator-spectroheliometer, an extreme ultraviolet and X-ray telescope, an X-ray spectrographic telescope, and a chromospheric extreme ultraviolet spectrograph, all within the ATM. The MDA contained equipment for space manufacturing experiments and externally mounted earth resources cameras and experiments including a multispectral photography facility, an earth terrain camera, an infrared spectrometer, a multispectral scanner, a microwave radiometer/scatterometer and altimeter and an L-band microwave radiometer.

During the unmanned launch of the station, a meteroid/thermal shield ripped loose from the OWS, severing one of the two major solar 'wings,' or power-generating panels. Temperatures inside the station were raised to uninhabitable levels and electrical power was cut drastically.

The first crew of astronauts arrived at Skylab aboard an Apollo spacecraft on 25 May 1973, 10 days after it was put into space. They were able to repair the durable space station so that it would be habitable not only for them but for the two Skylab crews that followed them. The first crew to man Skylab, Charles Conrad, Joseph Kerwin and Paul Weitz, lived in the station for 28 days. The second group, Alan Bean, Owen Garriott and Jack Lousma, arrived on 28 July 1973 and stayed for nearly 60 days; Gerald Carr, Edward Gibson and William Pogue arrived on 16 November 1973 and stayed for 84 days.

The designation system for the four Skylab program flights ran simply Skylab 1, 2, 3 and 4. The Skylab 1 launch was that of the Skylab space station itself, while the other three were actually Apollo launches whose destination was the Skylab station.

Skylab supported crews for a total of 172 days in relative comfort between May 1973 and February 1974. With the third Apollo flight to Skylab, the first phase of the program was concluded. It was intended that Skylab be reoccupied when the space shuttle program got underway. Gradually, however, the space station's orbit began to decay due to sun-spot activity during 1978-79 that caused the earth's atmosphere to expand. A space shuttle mission was planned for February 1980 in which astronauts would attach an inertial upper stage (IUS) to Skylab and boost it into higher orbit, where it could be reused. During that period, however, there was no American manned space capability and nothing could be done to save the space station.

On 11 July 1979, a year before the planned rescue mission and two years before the shuttle's actual first flight, Skylab plunged into the atmosphere and burned up over the Indian Ocean, with a few pieces of debris reaching ground in Australia. It was the last artifact of America's first period of manned space exploration.

(See Apollo/Skylab)

Skylab Space Station

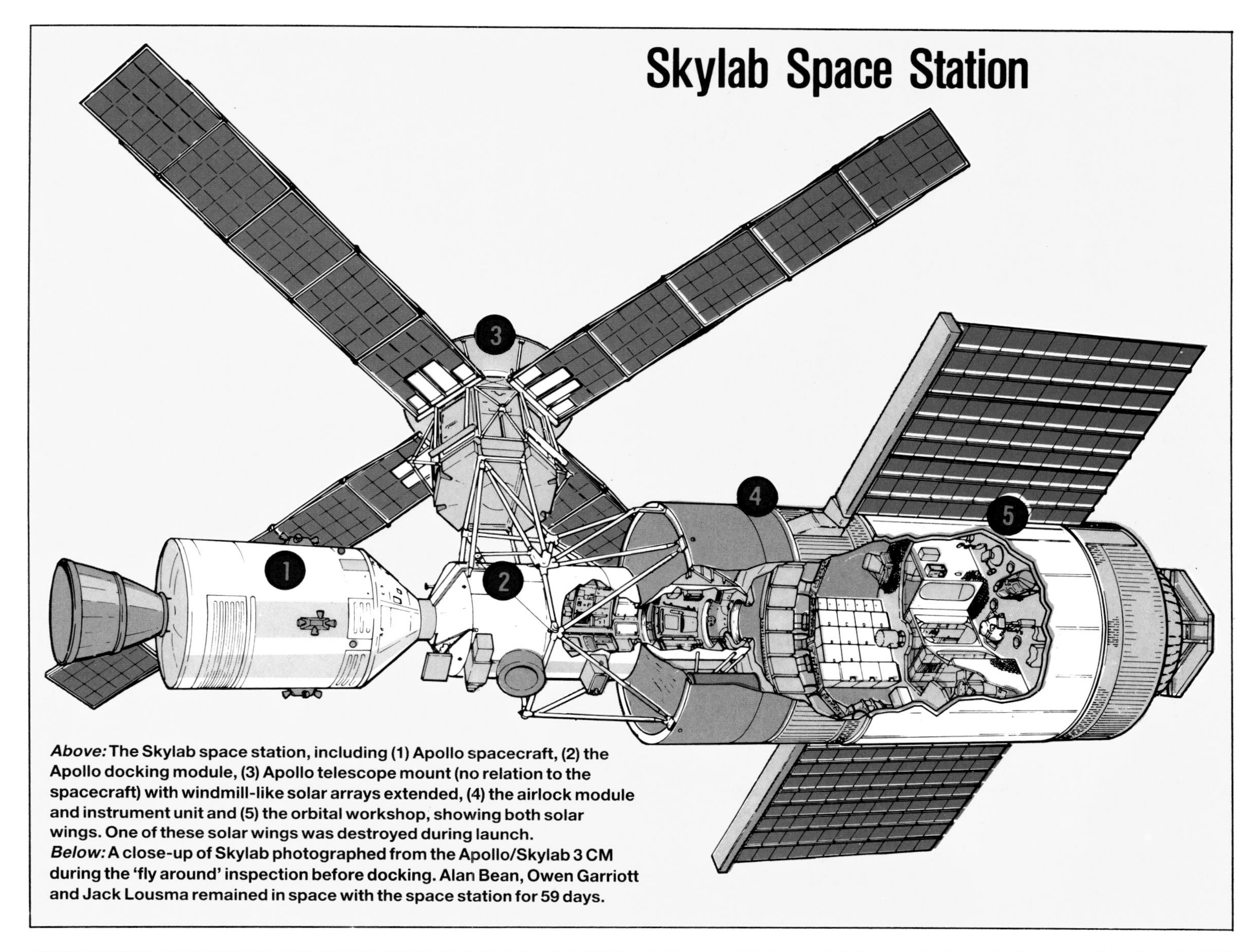

Above: The Skylab space station, including (1) Apollo spacecraft, (2) the Apollo docking module, (3) Apollo telescope mount (no relation to the spacecraft) with windmill-like solar arrays extended, (4) the airlock module and instrument unit and (5) the orbital workshop, showing both solar wings. One of these solar wings was destroyed during launch.
Below: A close-up of Skylab photographed from the Apollo/Skylab 3 CM during the 'fly around' inspection before docking. Alan Bean, Owen Garriott and Jack Lousma remained in space with the space station for 59 days.

Skylab Component	*Weight* (pounds)	*Diameter* (feet)	*Length* (feet)
Multiple docking adapter (MDA)	13,600	10	17.3
Apollo telescope mount (ATM)	24,800	10	13.3
Airlock module (AM)	49,200	10	17.6
Instrument unit (IU)	4,500	21.6	3
Orbital workshop (OWS)	76,000	21.6	48.1

Skylab
(Data is given for the Skylab space station, or the Skylab 1 mission; for Skylab 2, 3 and 4 flights see Apollo/Skylab.)

Launch vehicle: Saturn 5
Launch: 14 May 1973
Re-entry: 11 July 1979
Actual total weight: 167,849 pounds
Diameter: 21.6 feet (at widest point)
Length: 84.2 feet (excluding CSM)

UNITED

Appendix 2

Launch Vehicles

The role of launch vehicles is, quite simply, to put spacecraft into space. They are the booster rockets whose brute force permits the spacecraft to escape the earth's atmosphere. While spacecraft are generally (though not always) small and delicate mechanisms, launch vehicles are enormous and powerful engines.

The development of rockets can be traced back to the twelfth century, but it wasn't until 1926 that Robert Goddard launched in Auburn, Massachusetts the world's first liquid-fuel rocket, the forerunner of modern launch vehicles. Over the next two decades, rocket development continued on both sides of the Atlantic, with experiments in Britain, France, Germany and the Soviet Union. During World War II, German scientists, notably Dr Wernher von Braun, developed the A-4 single-stage rocket that was used as an offensive weapon with an explosive warhead under the designation V-2 (Vergence Weapon, second). On 3 October 1942 an A-4 was launched that reached into the fringes of outer space, to an altitude of 53 miles.

Plans to use the A-4 to explore near-earth space and even to launch a spacecraft were pre-empted by the war effort and such activities remained untried when Germany lost the war. Both the United States and Soviet Union confiscated A-4s that they found in Germany, and this single rocket type became the basis for postwar missile development in both countries.

The first major development in postwar rocket technology was the concept of a multistage rocket. The Germans had such a rocket, A-9/A-10, on the drawing boards in 1945, but it was the Americans two years later who first successfully launched a multistage vehicle. The idea behind a multistage rocket is that as the first stage reaches its apogee, say at 50 miles, the

Left: **Prelaunch of the Atlas-Agena vehicle with the nuclear-detection satellite Vela 3.** ***Right:*** **The first stage of the Saturn 5 dwarfed the entire Titan 2 used for the Gemini spacecraft, including Gemini 5 (*above*), on 21 August 1965.**

Above: **Vanguard TV-1 in the gantry, 1957.**
Right: **Gemini 9A was launched by Titan after two delays.** ***Left:*** **The Mercury Atlas was first used for manned spaceflight in February 1962.**

second stage is 'launched' from the first stage 50 miles closer to outer space. This principle is used today on all major launch vehicles, with three- and four-stage vehicles not uncommon. The US Army conducted a dozen two-stage tests in the late 1940s, using an A-4 first stage with a WAC-Corporal second stage. The upper stages reached a record altitude of 244 miles in February 1949, and carried scientific instrumentation, but the US space program was still nearly a decade away from placing a spacecraft into outer space.

The first American launch vehicle program was the US Navy's Vanguard, first announced in 1955. Fraught with problems, its progress was overtaken by the Soviet Vostok, which was used in October 1957 to launch Sputnik 1, the world's first spacecraft, and by the US Army's Juno 1, which in January 1958 launched Explorer 1, America's first spacecraft. The Vanguard program was abandoned as a failure, while the Juno evolved into the Redstone program, which resulted in the Redstone launch vehicle that was used to place the first US astronauts into space.

Meanwhile, the same rockets that the US Air Force was developing as ICBMs were being adapted as spacecraft launch vehicles. This included the General Dynamics/Convair Atlas (first used in

Above: The 'Delta Straight 8' launch vehicle put Landsat 2 into a near-perfect circular orbit.

Right: Saturn 5, the largest US launch vehicle ever, on its way to Pad 39 for launch.

1959), the Douglas Thor (1959) and the Martin Marietta Titan (1964). The Atlas was a major early vehicle and was used to launch the bulk of the Mercury manned spacecraft. Thor, the first US ballistic missile, gradually evolved into the long-tank Thor, which became the Thor Delta (Delta 1904), and then into the highly versatile Delta series of vehicles, which have been the backbone of NASA's middleweight launch capability since the late 1960s. The Titan, which served the US Air Force in its original ICBM form well into the 1980s, first served as a NASA launch vehicle in 1964 with the start of the Gemini manned spacecraft program. The Titan II that served the Gemini program evolved into the Titan III. In its various subtypes it serves both NASA and the US Air Force as a heavyweight launch vehicle for deep space and planetary probes and for particularly heavy loads such as the Air Force's enormous reconnaissance spacecraft.

While the Martin Titan is the US heavyweight launch vehicle and the McDonnell Douglas Delta the middleweight, the Vought Scout is the lightweight launch vehicle. Continually evolving since its first flights in 1960, Scout was designed to provide NASA and DOD with a relatively simple, low-cost means of orbiting light spacecraft and conducting re-entry tests.

In addition to these launch vehicles, NASA and DOD developed a number of specialized upper stages. The most important of these are the Lockheed Agena and General Dynamics/Convair Centaur. The versatile Agena has been used on more space missions than any other upper stage in the world. It has the capability of being used with a wide variety of basic launch vehicles from Thor to Titan, and was at one time an important part of the Atlas (Atlas Agena) program, though it is no longer used with Atlas. Agena can carry a variety of types and sizes of spacecraft, and its restartable engine means that it can be left attached to spacecraft and be used to boost them to different orbits. Among its notable achievements was its use as a docking target for manned Gemini spacecraft like Agena. Centaur has been used for a variety of purposes, including very long-range missions and those into deep space, such as the Viking and Voyager planetary probes, when it was teamed up with Titan 3 as Titan 3 Centaur.

When they were developed, the Saturn series of launch vehicles were the most powerful built by the US. The Saturn 5, which still dwarfs all other American boosters, was the largest and most power-

USA
UNITED STATES
USA
USA

Above: **The Scout launch vehicle is one of NASA's most versatile.** ***Left:*** **The Atlas Centaur launches the Surveyor 3 spacecraft.**

ful US launch vehicle ever built. Saturn was developed expressly for the Apollo spacecraft. Saturn 1B was used for Apollo developmental flights and Saturn 5 for actual manned flights.

The program began with the Saturn 1 (1A), which was used for 10 launches between October 1961 and July 1965 and achieved a then unprecedented, 100 percent success rate. These launch vehicles were used to place both satellites and unmanned boiler-plate test replicas of the Apollo spacecraft into earth orbit.

The Saturn 1 program was followed by the Saturn 1B program, which utilized upgraded elements of the Saturn 1 vehicle and offered 50 percent greater payload capacity than the Saturn 1. The Saturn 1B had a Chrysler first stage and a Douglas S-4B upper stage, and was first used on 26 February 1966 to place an unmanned Apollo test capsule into orbit. The Saturn 1B was used throughout the early Apollo tests and for the first manned Apollo mission (Apollo 7) in October 1968. Overshadowed by the Saturn 5 through most of the Apollo program, the Saturn 1B made a comeback in 1973 as the launch vehicle that took all three Apollo/Skylab crews to the Skylab space station. The ninth and last flight of the 100 percent successful Saturn 1B, and of the whole Saturn program, came in July 1975 with the launch of the Apollo component of the historic US-USSR Apollo-Soyuz rendezvous in space.

The Saturn 5, successor to the Saturn 1B launch vehicle, was the vehicle used for all of the Apollo lunar missions. It was much larger than the Saturn 1B by twice the height and six times the weight. Boeing was the prime contractor for the first stage and Rockwell for the second. McDonnell Douglas supplied the S-4B third stage, which was basically the same as the S-4B used in the Saturn 1B. Development of the Saturn 5 began in January 1962, and it was first used on 9 November 1967 to launch the unmanned Apollo 4 test capsule. A total of 15 Saturn 5s were produced, with 12 used for 13 successful missions, including all Apollo

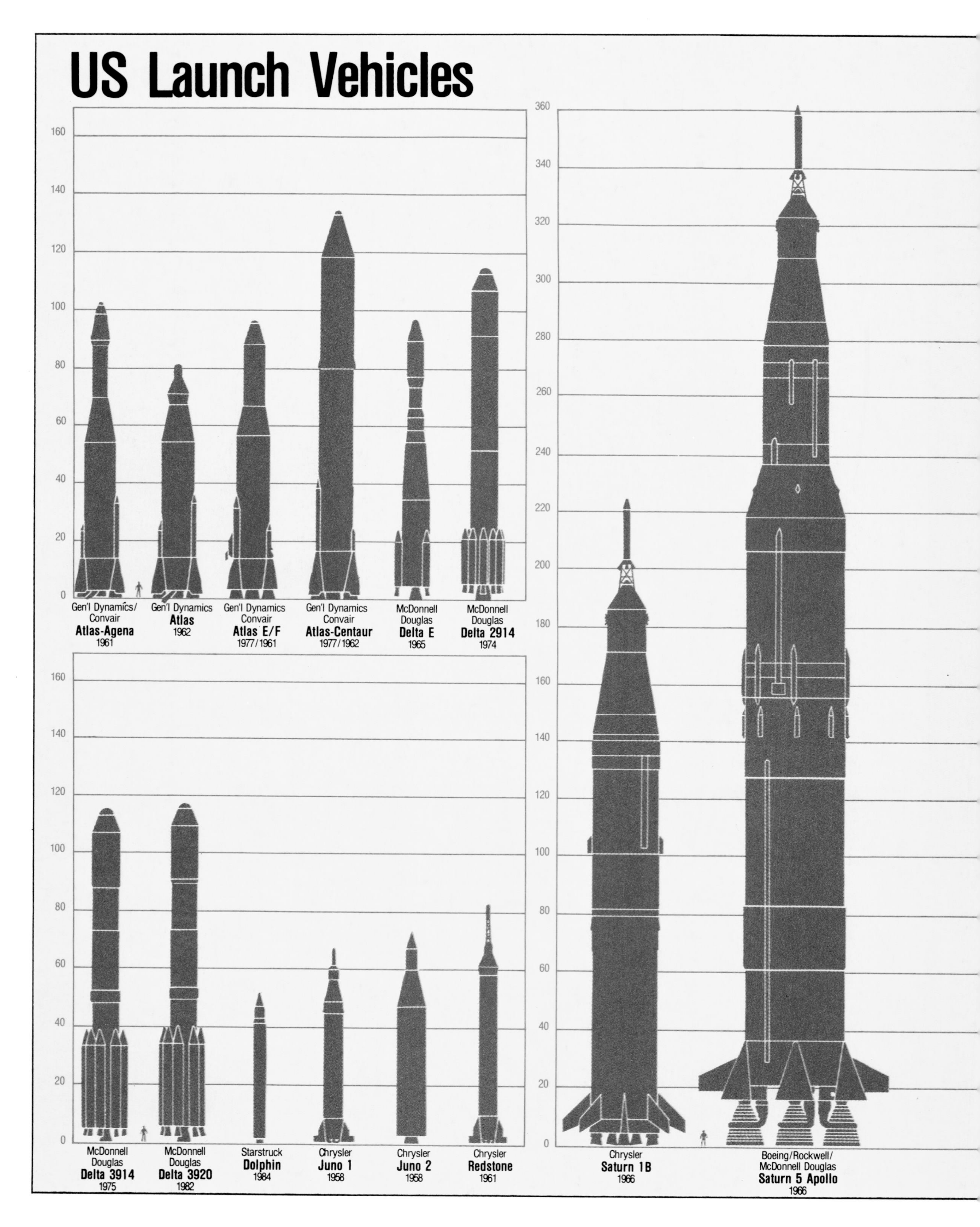

US Launch Vehicles
Gen'l Dynamics/ Convair
Atlas-Agena
1961
Gen'l Dynamics
Atlas
1962
Gen'l Dynamics Convair
Atlas E/F
1977/1961
Gen'l Dynamics Convair
Atlas-Centaur
1977/1962
McDonnell Douglas
Delta E
1965
McDonnell Douglas
Delta 2914
1974
McDonnell Douglas
Delta 3914
1975
McDonnell Douglas
Delta 3920
1982
Starstruck
Dolphin
1984
Chrysler
Juno 1
1958
Chrysler
Juno 2
1958
Chrysler
Redstone
1961
Chrysler
Saturn 1B
1966
Boeing/Rockwell/ McDonnell Douglas
Saturn 5 Apollo
1966

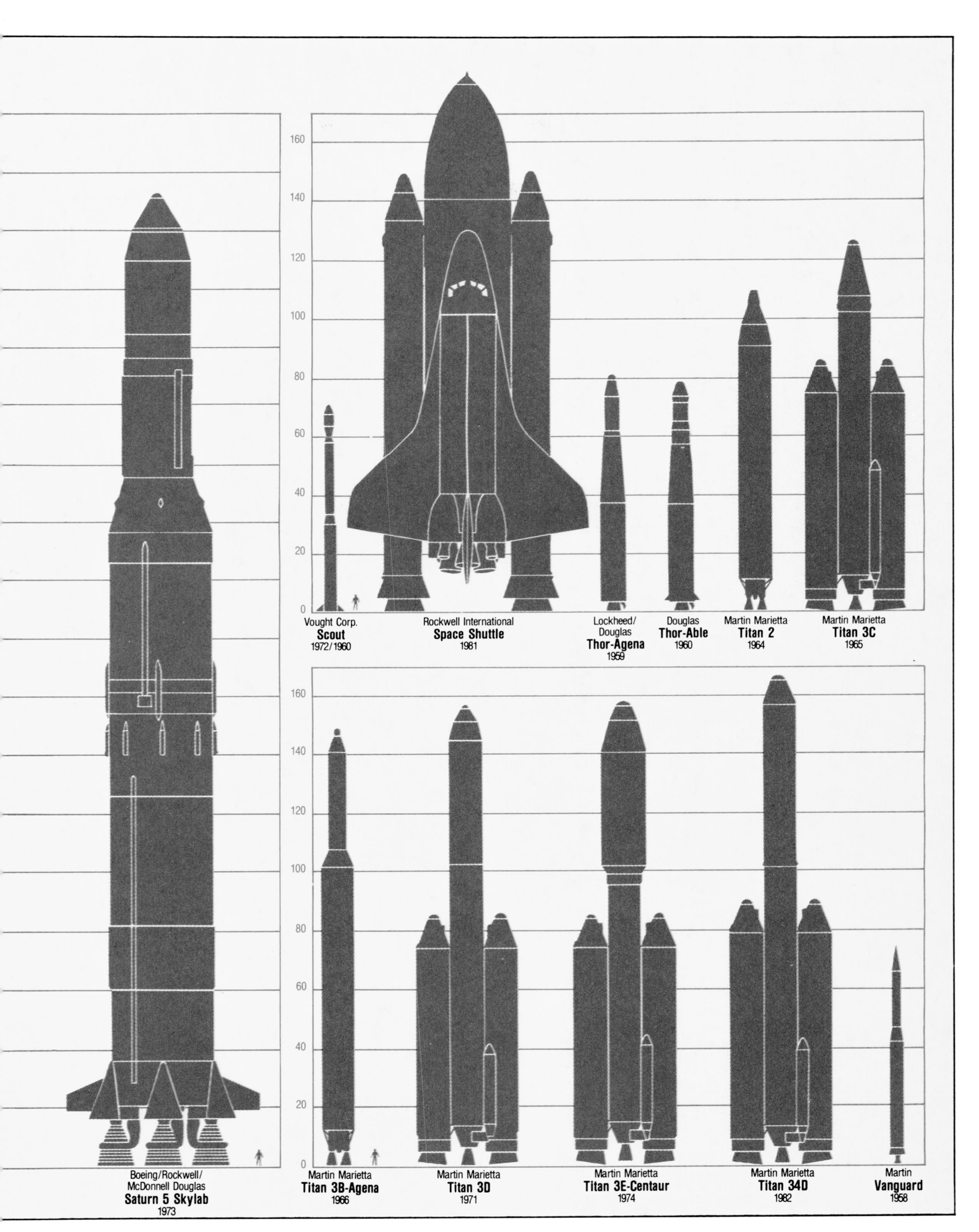
0
20
40
60
80
100
120
140
160
Vought Corp.
Scout
1972/1960
Rockwell International
Space Shuttle
1981
Lockheed/
Douglas
Thor-Agena
1959
Douglas
Thor-Able
1960
Martin Marietta
Titan 2
1964
Martin Marietta
Titan 3C
1965
Boeing/Rockwell/
McDonnell Douglas
Saturn 5 Skylab
1973
Martin Marietta
Titan 3B-Agena
1966
Martin Marietta
Titan 3D
1971
Martin Marietta
Titan 3E-Centaur
1974
Martin Marietta
Titan 34D
1982
Martin
Vanguard
1958

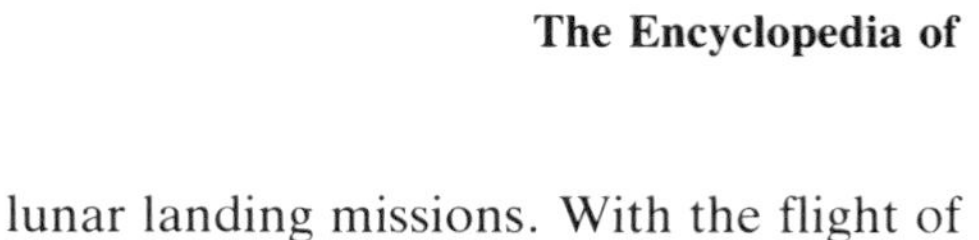

lunar landing missions. With the flight of the last Apollo spacecraft in 1975, the remaining Saturn launch vehicles were placed in permanent storage.

The space shuttle, or Space Transportation System (STS), has as one of its integral parts, the only major US launch vehicle designed from the ground up since the Saturn program. The STS program consists of three major components, which include the Rockwell orbiter, the Martin Marietta propellant/ballast tank and the two Thiokol solid-propellant rocket boosters. These boosters, constituting the launch vehicle portion of the

Left: **The Intelsat 4 spacecraft atop an Atlas Centaur, August 1973.** ***Right:*** **The Saturn 5 launch vehicle generated a 7.7-million-pound lift-off thrust.** ***Below:*** **A Titan 3 with Centaur upper stage sends Viking 2 toward its rendezvous with Mars.**

STS, deliver 3,239,600 pounds of thrust at lift-off. This is roughly double the thrust of the Saturn 1B and 28 percent more than the thrust of the most powerful Titan. Only the enormous Saturn 5, with its maximum 7,723,726 pounds of thrust, was more powerful (see also Space Shuttle).

At the other end of the spectrum from the powerful STS boosters are a number of very small launch vehicles that are being developed entirely by private industry, with no NASA or DOD fiscal involvement. While all American launch vehicles (unlike many US spacecraft) are built by private firms, they have all been built under government contract. For example, RCA may wish to launch its Satcom spacecraft, and the launch may take place aboard a McDonnell Douglas Delta launch vehicle, but the Delta is launched by NASA. Until the 1980s, all US spacecraft were launched and managed by the government (DOD or NASA), even though the spacecraft were owned by private companies and operated for commercial purposes.

By the late 1970s, a number of entrepreneurs decided that they could offer a cost-effective alternative for satellite users. Several small firms, mostly in the San Francisco Bay Area, began to develop launch vehicles that they could build and launch at a fraction of the cost charged by NASA. Among these firms are Pacific American Launch Systems, Truax Engineering and Starstruck, Incorporated. A major milestone in the evolution of privately built/privately financed launch vehicles came with the successful launch of Starstruck's Dolphin launch vehicle from a site off the California coast on 3 August 1984.

US Launch Vehicles

Vehicle	Contractor	Date	Length (ft)	Maximum Diameter (ft)	Total Weight (lb)	Stage	Stage Contractor	Propellant	Pounds thrust
Atlas (Score)	General Dynamics/ Convair	1958	85.0	10	244,000	0	Rocketdyne LR 89-NA5 (×2)	LO_2/RP-1	330,000
				16 (skirt)		1	Rocketdyne LR 105-NA5 + LR 101 verniers (×2)	LO_2/RP-1	59,000
Atlas-Able	General Dynamics/ Convair	1959	98.0	10	260,000	0	Rocketdyne LR 89-NA5 (×2)	LO_2/RP-1	330,000
				16		1	Rocketdyne LR 105-NA5 + LR 101-5 verniers (×2)	LO_2/RP-1	59,000
						2	AJ10-11B	IWFNA/UDMH	7,500
						3	ABL X-248-3 Altair	Solid	3,100
Atlas-Agena A	General Dynamics/ Convair, USAF	1960	99.0	10	273,000	0	Rocketdyne LR 89-NA5 (×2)	LO_2/RP-1	330,000
				16		1	Rocketdyne LR 105-NA5 + 101-7 verniers (×2)	LO_2/RP-1	59,000
						2	Bell 8096	IRFNA/UDMH	15,000
Atlas-Agena B	General Dynamics/ Convair, USAF, NASA	1961	98.0	10	275,000	0	Rocketdyne LR 89-NA5 (×2)	LO_2/RP-1	330,000
				16		1	Rocketdyne LR 105-NA5 + 101-7 verniers (×2)	LO_2/RP-1	59,000
						2	Bell 8096	IRFNA/UDMH	15,000
(Mercury) Atlas	General Dynamics/ NASA, MSFC	1962	95.3	10.0	260,000	0	Rocketdyne LR 89-NA5 (×2)	LO_2/RP-1	367,000
						1	Rocketdyne LR 105-NA5 + LR 101-NA7 verniers	LO_2/RP-1	59,000
Atlas-Agena D	General Dynamics/ Convair, USAF, NASA	1963	102.0	10	274,965	0	Rocketdyne LR 89-NA5 (×2)	LO_2/RP-1	330,000
				16		1	Rocketdyne LR 101-NA5 + LR 101 verniers (×2)	LO_2/RP-1	59,000
						2	Bell 8096	IRFNA/UDMH	16,000
Atlas-Burner 2	General Dynamics/ Boeing, USAF	1968	78.7	10	286,440	0	Rocketdyne YLR 89-NA7 (×2)	LO_2/RP-1	370,000
				16		1	Rocketdyne YLR 105-NA7 + LR 101 verniers (×2)	LO_2/RP-1	59,000
						2	Thiokol TE-M-364-2	Solid	8,800
Atlas-Centaur	General Dynamics/ Convair	1962	105.0	10	325,732	0	Rocketdyne YR 89-NA5 (×2)	LO_2/RP-1	367,000
				16		1	Rocketdyne LR 105-NA5 + LR 101 verniers (×2)	LO_2/RP-1	59,000
						2	P&W RL 10A-3-3 (×2)	LO_2/LH_2	30,000
Atlas-Centaur	General Dynamics/ Convair, NASA	1966	131.0	10	329,017	0	Rocketdyne YLR 89-NA7 (×2)	LO_2/RP-1	370,000
				16		1	Rocketdyne YLR 105-NA7 + LR 101 verniers (×2)	LO_2/RP-1	61,300
						2	P&W RL 10A-3-3 (×2)	LO_2/LH_2	30,000
Atlas F	General Dynamics/ Convair, USAF, NASA	1977	85.0	10	262,500	0	Rocketdyne LR 89-NA5 (×2)	LO_2/RP-1	330,000
				16		1	Rocketdyne LR 105-NA7 + LR 101 verniers (×2)	LO_2/RP-1	59,000
						2	TE-364-4	Solid	14,330
Delta B DSV-3B	McDonnell Douglas, NASA	1962	87.9	8.0	114,170	1	Rocketdyne LR 79-NA11	LO_2/RJ-1	172,000
						2	Aerojet AJ-10-118D	IRFNA/UDMH	7,575
						3	X-248 A5DM	Solid	2,760
Delta D DSV-3D	McDonnell Douglas, NASA	1964	92.8	8.0	143,325	0	Thiokol Castor 1 (×3)	Solid	156,775
						1	Rocketdyne LR 79-NA13	LO_2/RJ-1	175,075
						2	Aerojet AJ-10-118D	IRFNA/UDMH	7,715
						3	Hercules -258	Solid	2,755
Delta E DSV-3E	McDonnell Douglas, NASA	1965	95.8	8.0	149,940	0	Thiokol Castor 1 (×3)	Solid	156,775
						1	Rocketdyne LR 79-NA13	LO_2/RJ-1	175,075
						2	Aerojet AJ-10-118E	IRFNA/UDMH	7,805
						3	Hercules X-258	Solid	2,755
Delta 1000 (1913)	McDonnell Douglas, NASA	1972	116.0	8.0	295,470	0	Thiokol Castor 2 (×9)	Solid	470,325
						1	Rocketdyne LR 79-NA13	LO_2/RJ-1	175,075
						2	Aerojet AJ-10-118F	N_2O_4/Aerozine 50	9,810
						3	Thiokol TE-364-3	Solid	9,505

US Launch Vehicles

Vehicle	Contractor	Date	Length (ft)	Maximum Diameter (ft)	Total Weight (lb)	Stage	Stage Contractor	Propellant	Pounds thrust
Delta 2000 (2914)	McDonnell Douglas, NASA	1974	116.0	8.0	293,110	0	Thiokol Caster 2 (×9)	Solid	470,325
						1	Rocketdyne RS-27	LO_2/RP-1	205,065
						2	TRW TR-201	N_2O_4/Aerozine 50	9,810
						3	Thiokol TE-364-4	Solid	14,995
Delta 3914	McDonnell Douglas, NASA	1975	116.0	8.0	420,269	0	Thiokol Caster 4 (×9)	Solid	766,015
						1	Rocketdyne RS-27	LO_2/RP-1	205,065
						2	TRW TR-201	N_2O_4/Aerozine 50	9,810
						3	Thiokol TE-364-4	Solid	14,995
Delta 3910	McDonnell Douglas, NASA	1980	116.0	8.0	418,000	0	Thiokol Castor 4 (×9)	Solid	766,015
						1	Rocketdyne RS-27	LO_2/RP-1	205,065
						2	TRW TR-201	N_2O_4/Aerozine 50	9,810
Delta 3910/PAM	McDonnell Douglas, NASA	1980	116.0	8.0	422,100	0	Thiokol Caster 4 (×9)	Solid	766,015
						1	Rocketdyne RS-27	LO_2/RP-1	205,065
						2	TRW TR-201	N_2O_4/Aerozine 50	9,810
						3	Thiokol STAR 48	Solid	18,500
Delta 3910/PAM	McDonnell Douglas, NASA	1980	116.0	8.0	422,100	0	Thiokol Caster 4 (×9)	Solid	766,015
						1	Rocketdyne RS-27	LO_2/RP-1	205,065
						2	TRW TR-201	N_2O_4/Aerozine 50	9,810
						3	Thiokol STAR 48	Solid	18,500
Delta 3920/PAM	McDonnell Douglas, NASA	1982	116.0	8.0	426,000	0	Thiokol Center 4 (×9)	Solid	766,015
						1	Rocketdyne RS-27	LO_2/RP-1	205,065
						2	Aerojet AJ10-118K-1TR1	N_2O_4/Aerozine 50	9,810
						3	Thiokol STAR 48	Solid	18,500
Dolphin	Starstruck	1984	51.0	3.5	18,000	1	Starstruck PTS-1 D-1 Engine	Hybrid Polybuladyne rubber/ LO_2	42,000
Juno 1	Chrysler, NASA, MSFC	1958	71.25	5.8	64,000	1	Rocketdyne A-7	LO_2/'Hydyne'	83,000
						2	Thiokol Sergeant (mod) (×11)	Solid	16,500
						3	Thiokol Sergeant (mod) (×3)	Solid	5,400
						4	Thikol Sergeant (mod)	Solid	1,800
Juno 2	Chrysler, NASA, MSFC	1958	76.6	8.75	122,000	1	Rocketdyne S-3D	LO_2/RP-1	150,000
						2	Thiokol Sergeant (mod) (×11)	Solid	16,500
						3	Thiokol Sergeant (mod) (×3)	Solid	5,400
						4	Thikol Sergeant (mod)	Solid	1,800
Mercury-Redstone	Chrysler, NASA, MSFC	1961	83.0	5.8	66,000	1	Rocketdyne A-7	LO_2/ethyl alcohol + water	78,000
Saturn 1	Chrysler, NASA	1961	164.0	21.5	1,122,000	1	Rocketdyne H-1 (×8)	LO_2/RP-1	1,504,000
						2	P&W RL-10A-3	LO_2/LH_2	90,000
Saturn 1B	Chrysler, NASA, MSFC	1966	224.0	21.7	1,295,000	1	Rocketdyne H-1 (×8)	LO_2/RP-1	1,640,000
						2	Rocketdyne J-2	LO_2/LH_2	225,000
Saturn 5/Apollo	NASA, MSFC, Boeing/Rockwell, McDonnell Douglas	1967	363.0	33.0	6,423,000	1	Rocketdyne F-1 (×5)	LO_2/RP-1	7,650,000
						2	Rocketdyne J-2 (×5)	LO_2/LH_2	1,150,000
						3	Rocketdyne J-2	LO_2/LH_2	238,000
Saturn 5/Skylab	NASA, MSFC, Boeing/Rockwell	1973	333.7	33.0	6,222,000	1	Rocketdyne F-1 (×5)	LO_2/RP-1	7,724,000
						2	Rocketdyne J-2 (×5)	LO_2/LH_2	1,125,000
Scout	Vought Corp, NASA	1960	72.0	3.3	36,600	1	UTC Algol I	Solid	115,000
						2	Thiokol Castor I	Solid	50,000
						3	ABL X-254 Antares I	Solid	13,000
						4	ABL X-248 Altair I	Solid	3,000

Note: The numeral '0' listed under *Stage* indicates 'strap-on' engines, or supplementary engines attached to the first stage.

US Launch Vehicles (contd)

Vehicle	Contractor	Date	Length (ft)	Maximum Diameter (ft)	Total Weight (lb)	Stage	Stage Contractor	Propellant	Pounds thrust
Scout D	Vought Corp, NASA, USAF	1972	75.5	3.8	47,000	1	UTC Algol IIIA	Solid	108,000
						2	Thiokol Castor IIA	Solid	63,000
						3	H X-258 Antares IIA	Solid	28,500
						4	Thiokol Altair 3A	Solid	5,900
Space Shuttle	Rockwell International, NASA	1981	184.2	78.1 (wing span)	4,500,000	0	Thiokol SRB (×2)	Solid	5,300,000
			122.2			1	Rocketdyne SSME (×3) +	LO_2/LH_2	1,410,000
							Aerojet OMS (×2)	N_2O_4/MMH	12,000
Thor-Able	Douglas, NASA, USAF	1958	90.0	8.0	114,660	1	Rocketdyne LR 79-NA9	LO_2/RJ-1	150,000
						2	Aerojet AJ-10-42	IRFNA/UDMH	7,575
						3	ABL X-248 Altair	Solid	2,760
Thor-Able Star	Douglas, USAF	1960	79.3	8.0	117,900	1	Rocketdyne LR 79-NA11	LO_2/RJ-1	172,000
						2	Aerojet AJ-10-104 with restart capability	IR (or W) FNA/UDMH	7,730
Thor-Agena A	Lockheed/ Douglas, USAF	1959	78.5	8.0	117,000	1	Rocketdyne XLR 79-NA9	LO_2/RJ-1	150,000
						2	Bell Hustler 8048	IRFNA/UDMH	15,500
Thor-Agena B	Lockheed/ Douglas, USAF	1960	81.3	8.0	123,040	1	Rocketdyne LR 79-NA11	LO_2/RJ-1	172,000
						2	Bell Hustler 8096	IR (or W) FNA/UDMH	16,000
Thor-Agena D	Lockheed/ Douglas, USAF, NASA	1962	76.3	8.0	123,040	1	Rocketdyne XLR 79-NA11	LO_2/RJ-1	172,000
						2	Bell 8096	IRFNA/UDMH	16,000
Thor-Burner 2A	Lockheed/ McDonnell Douglas, Boeing, USAF	1971	85.0	8.0	N/A	1	Rocketdyne XLR 79-NA11	LO_2/RJ-1	172,000
						2	Thiokol TE-M-364-2	Solid	10,000av
						3	Thiokol TE-M-442-1	Solid	8,800
Titan IIIA	Martin Marietta, USAF	1964	108.0	10.0	407,925	1	Aerojet LR-87 (×2)	N_2O_4/Aerozine 50	430,000
						2	Aerojet LR-91	N_2O_4/Aerozine 50	100,000
						3	Transtage	N_2O_4/Aerozine 50	16,000
(Gemini) Titan II	Martin Marietta, NASA	1964	109.0	10.0	407,925	1	Aerojet LR-87 (×2)	N_2O_4/Aerozine 50	430,000
						2	Aerojet LR-91	N_2O_4/Aerozine 50	100,000
Titan IIIB-Agena	Martin Marietta, USAF	1966	160.0 max	10.0	454,450	1	Aerojet LR-87-AJ-11 (×2)	N_2O_4/Aerozine 50	463,200
						2	Aerojet LR-91-AJ-11	N_2O_4/Aerozine 50	101,000
						3	Agena	IRFNA/UDMH	16,800
Titan IIIC	Martin Marietta, USAF, NASA	1965	157.0 max	10.0	1,392,000	0	UA 1205 (×2)	Solid	2,360,000
						1	Aerojet LR-87-AJ-11	N_2O_4/Aerozine 50	532,000
						2	Aerojet LR-91-AJ-11	N_2O_4/Aerozine 50	101,000
						3	Trans AJ10-138	N_2O_4/Aerozine 50	16,000
Titan IIID	Martin Marietta, USAF, NASA	1971	155.0 max	10.0	1,300,000	0	UA 1205 (×2)	Solid	2,360,000
						1	Aerojet LR-870-AJ-11	N_2O_4/Aerozine 50	532,000
						2	Aerojet LR-87-AJ-11	N_2O_4/Aerozine 50	101,000
Titan IIIE-Centaur	Martin Marietta, NASA	1974	160.0	10.0	1,411,200	0	UA 1205 (×2)	Solid	2,361,000
						1	Aerojet YLR-870-AJ-11	N_2O_4/Aerozine 50	530,000
						2	Aerojet YLR-910-AJ-11	N_2O_4/Aerozine 50	101,000
						3	Centaur D-1T	LO_2/LH_2	30,000
Titan 34D	Martin Marietta, NASA, USAF	1982	160.7 max	10.0	1,742,530	0	UA 1205 (×2)	Solid	2,498,000
						1	Aerojet LR-87-OAJ-11	N_2O_4/Aerozine 50	530,000
						2	Aerojet LR-91-OAJ-11	N_2O_4/Aerozine 50	101,000
						3	IUS Stage 1	Solid	62,000
						4	IUS Stage 2	Solid	26,000
Vanguard	Glenn L. Martin, US Navy	1958	72.0	3.7	22,600	1	GE X-405	LO_2/Kerosene	28,000
						2	Aerojet AJ-10	IWFNA/UDMH	7,500
						3	ABL X-248 Altair	Solid	3,100

Note: The numeral '0' listed under *Stage* indicates 'strap-on' engines, or supplementary engines attached to the first stage.

Appendix 3

Acronyms and Abbreviations

The following list of acronyms and abbreviations includes those used in this book and those that the reader may encounter in other literature concerning US spacecraft and the US space program.

ABL Allegamy Ballistic Laboratory
ABM apogee boost motor
ABMA Army Ballistic Missile Agency
ABMDA Army Ballistic Missile Defense Agency
ACN ascension
ACS attitude control system
ADAU analog data acquisition unit
ADC Air Defense Command
ADCOM Air Defense Command
ADP automatic data processor
AE atmosphere explorer
AEC Atomic Energy Commission
AEM-A Applications Explorer Mission A
AFB Air Force base
AFGL Air Force Geophysical Laboratory
AFLC US Air Force Logistics Command
AFO announcement of flight opportunities
AFRPL US Air Force Rocket Propulsion Laboratory
AFSATCOM US Air Force Satellite Communications System
AFSC US Air Force Systems Command
AGE automated ground equipment
AGO Santiago, Chile
ALMV Air Launched Miniature Vehicle (ASAT)
ALRC Aerojet Liquid Rocket Company
ALSEP Apollo Lunar Surface Experiments Package
AM airlock module
AMPTE Active Magnetosphere Particle Tracer Explorer
AMSAT Amateur Satellite Corporation
AMU astronaut maneuvering unit (preceded MMU)
AOK All OK
AOP advanced onboard processor
AP automotive (automatic) picture transmission
APL Applied Physics Laboratory (Johns Hopkins University)
APT automatic picture transmission
ARC Ames Research Center
ARPA Advanced Research Projects Agency (later DARPA)
ASAT antisatellite weapon
ASPC Aerojet Solid Propulsion Company
ASS atmospheric sounding system
ASTP Apollo-Soyuz Test Project
AT&T American Telephone and Telegraph
ATDA Augmented Target Docking Adapter (Gemini)
ATM Apollo telescope mount
ATS Applications Technology Satellite
AU astronomical units
AVCS advanced vidicon camera system
AVHRR advanced very high resolution radiometer
BER bit error rate
BL Bell Laboratories
BMD Ballistic Missile Defense
BMEWS Ballistic Missile Early Warning System
BMO Ballistic Missile Office (USAF)
BUS backscatter ultraviolet spectrometer
C&DH Communications and Data Handling
C&W caution and warning
CAS Cooperative Applications Satellite
CC Cape Canaveral
CCC Central Command Control
CCE Change Compositive Explorer
CCIR International Radio Consultive Committee
CDT Central Daylight Time
CEPE Cylindrical Electrostatic Probe Experiment
CIA Central Intelligence Agency
CK Cape Kennedy (formerly Cape Canaveral)
CM command module
CMM Command Memory Management
CNES French National Center for Space Studies
COBE Cosmic Background Explorer
COMSAT Communications Satellite Corporation
CP communications processor
CPU central processing unit
CRL Cambridge Research Laboratory (USAF)
CRT cathode ray tube
CSM command plus service module unit (Apollo)
CST Central Standard Time
CSU cross-strap unit
CTC central telemetry control
CTS Communications Technology Satellite (Canada)
CYI Canary Islands
CZCS coastal zone color scanner
DARPA Defense Advanced Research Projects Agency (formerly ARPA)
DDAU digital data acquisition unit
DDPS Digital Data Processing System
DET direct energy transfer
DFVLR German Aerospace Ministry
DM docking module
DMA direct memory access
DMSP Defense Meteorological Satellite Program
DOD Department of Defense
DPDT double-pole double-throw
DPF Data Processing Facility
DRL Data Reduction Laboratory
DSCS Defense Satellite Communications System
DSN Deep Space Network
DSPS Defense Support Program Satellite(s)
ECM electronic countermeasures
EDT Eastern Daylight Time
EED electrical explosive device
EGSE electrical ground support equipment
EKG electrocardiograph
ELINT electronic intelligence
EMC electromagnetic compatibility
EMI electromagnetic interference
EMP electromagnetic pulse effect
EODAP Earth and Ocean Dynamics Applications Program
EOS Earth Observation Satellite
ERBS earth radiation budget system (later, earth radiation budget satellite)
ERCS Emergency Rocket Communications System
ERTS Earth Resources Technology Satellite
ESA European Space Agency

ESMR	electrically scanning microwave radiometer
ESSA	Environmental Science Services Administration
EST	Eastern Standard Time
ET	external tank
ET	extraterrestrial
ETR	Eastern Test Range
ETU	Engineering Test Unit
EVA	extravehicular activity
EVM	earth-viewing module
EW	electronic warfare
FAA	Federal Aeronautics Admin
FLTSATCOM	Fleet Satellite Communications System (US Navy)
FMOC	Flight Maneuver Operations Center
FPR	flat plate radiometer
FRCI	fibrous refactory composite insulation
FS	factors of safety
FSS	Flight Support System
FWS	filter wedge spectrometer
FY	fiscal year
GATV	Gemini Agena target vehicle
GD	General Dynamics
GDS	Goldstone
GE	General Electric
GEOAP	ground equivalent onboard attitude processor
GEODSS	Ground-based Electro-Optical Deep-Space Surveillance
GEOS	Geodynamic Experimental Ocean Satellite
GET	Ground Elapsed Time
GHz	Gigahertz
GISS	Goddard Institute for Space Studies
GM	General Motors
GMI	Goddard Management Instruction
GMT	Greenwich Mean Time
GOES	Geostationary Operational Environmental Satellite
GPS	Global Positioning System
GSE	ground support equipment
GSFC	Goddard Space Flight Center
GTDS	Ground Tracking Data Station
GTE	General Telephone and Electronics
GWM	Guam
HAW	Hawaii
HCMM	Heat Capacity Mapping Mission
HEAO	High Energy Astronomy Observatory
HEO	high earth orbit
HRIR	high-resolution infrared radiometer
HT	history tape
I&T	integration and tape
IBM	International Business Machines
ICATS	Integrated Command and Telemetry System
ICBM	Intercontinental Ballistic Missile
IDC	image dissector camera
IDSCP	Initial Defense Satellite Communications Program
IGY	International Geophysical Year (1957-58)
IMEWS	Integrated Missile Early Warning Satellite
IMP	Interplanetary Monitoring Platform
I/O	input/output
IOC	initial operating capabilities
IPD	Information Processing Division
IQSY	International Quiet Sun Years (1964-65)
IR	infrared
IRA	inertial reference assembly
IRAC	Interdepartment Radio Advisory Committee
IRAS	Infrared Astronomical Satellite
IRB	Integrated Requirements Board
IRBM	Intermediate Range Ballistic Missiles
IRIS	infrared inferometer spectrometer
IRLS	interrogation, recording and location system
IRM	ion-release module
IRR	infrared radiometer
IS	Implementation Staff
ISEE	International Sun-Earth Explorer
ISIS	International Satellites for Ionospheric Studies
ITOS	Improved TIROS Operational System
ITPR	infrared temperature profile radiometer
ITSS	Integrated Tactical Surveillance System
ITT	International Telephone and Telegraph
IU	instrument unit
IUE	International Ultraviolet Explorer
IUS	inertial upper space
IWFN	inhibited white fuming nitric acid
JPL	Jet Propulsion Laboratory

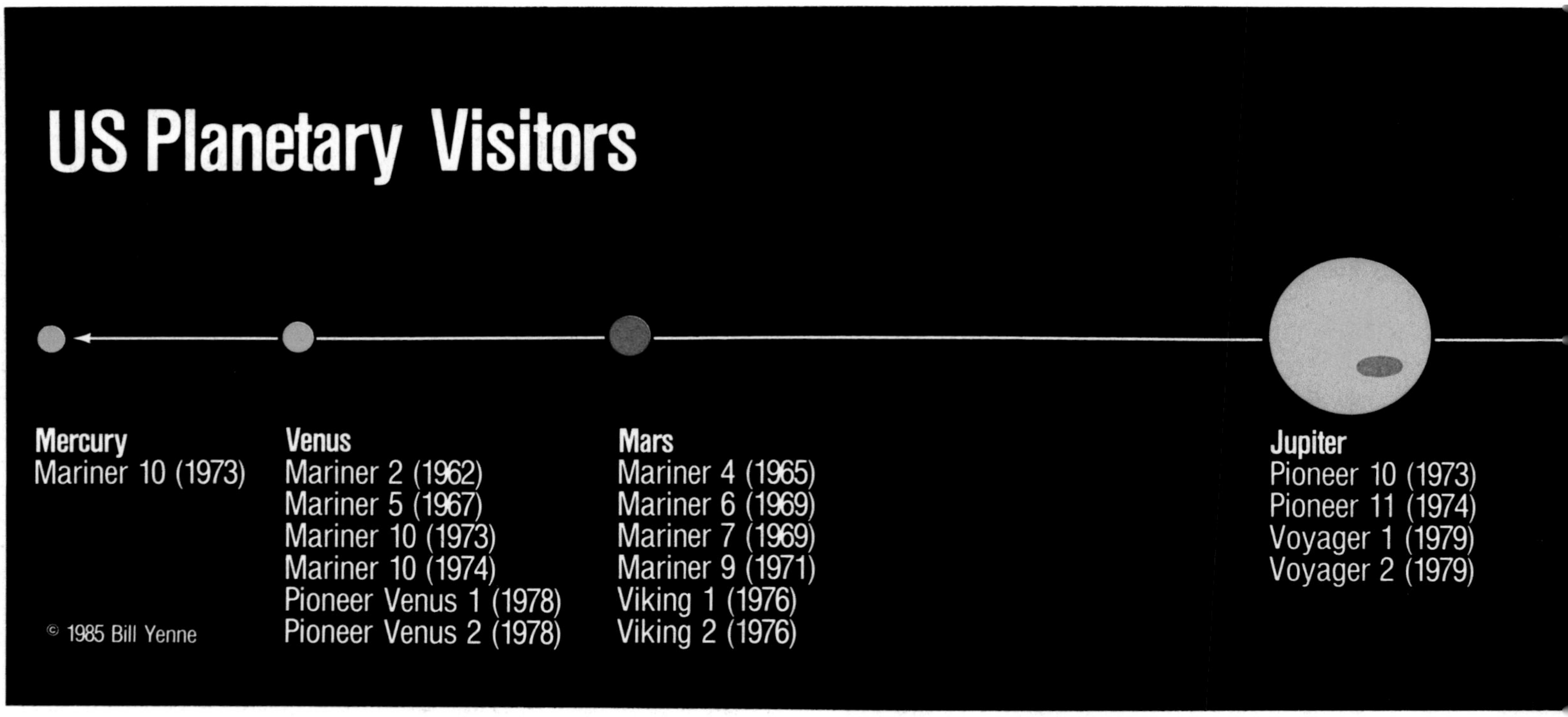

US Planetary Visitors

Mercury
Mariner 10 (1973)

Venus
Mariner 2 (1962)
Mariner 5 (1967)
Mariner 10 (1973)
Mariner 10 (1974)
Pioneer Venus 1 (1978)
Pioneer Venus 2 (1978)

Mars
Mariner 4 (1965)
Mariner 6 (1969)
Mariner 7 (1969)
Mariner 9 (1971)
Viking 1 (1976)
Viking 2 (1976)

Jupiter
Pioneer 10 (1973)
Pioneer 11 (1974)
Voyager 1 (1979)
Voyager 2 (1979)

© 1985 Bill Yenne

JSC Johnson Space Center
KH Key Hole
KHz Kilohertz (one million hertz)
KSC Kennedy Space Center (also known as Cape Canaveral, Cape Kennedy, Eastern Test Range, Merritt Island)
LAGEOS Laser GEOS
LARC Langley Research Center
LEM lunar excursion module (later LM)
LEO low earth orbit
LES launch escape system
LH_2 liquid hydrogen
LM landing module/lunar module
LO_2 liquid oxygen
LOX liquid oxygen
LRV lunar roving vehicle
LSIRR limb scanning infrared radiometer
LT launch time
LTA lunar module test article
LTOF Low-Temperature Optical Facility
LTV Ling Temco Vought
LV launch vehicle
Magsat Magnetic Field Satellite
MAM mission adapter module
MBB Messerschmitt Boelkow Blohm
MCC Mission Control Center
MDA multiple docking adapter
MDAC McDonnell Douglas Astronautics Company
MDT Mountain Daylight Time
ME main engine
MEM Module Exchange Magazine
MEMS Module Exchange Mechanism System
MFR multifunction receiver
MHz Megahertz (one million Hertz)
MIDAS Missile Defense Alarm System
MIL Merritt Island
MIR manipulated information rate
MIT Massachusetts Institute of Technology
MLRS Multiple Launch Rocket System
MMH mono-methyl hydrazine
MMS multimission modular spacecraft
MMU manned maneuvering unit
MOL Manned Orbital Laboratory
MR Mercury Redstone
MRB Management Review Board
MRIR medium-resolution infrared radiometer
MS Mercury Scout
MSAD Multisatellite Attitude Determination System
MSC Lockheed Missile and Space Company
MSC Manned Space Center, Houston, Texas
MSD Rockwell Missile Systems Division
MSFC Marshall Space Flight Center
MSFN Manned Space Flight Tracking Network
MSS multispectral scanner
MST Mountain Standard Time
MUSE monitor of ultraviolet solar energy
MWA momentum wheel assembly
MX missile, experimental
NAC Naval Avionics Center
NASA National Aeronautics and Space Administration
NASC Naval Air Systems Command
NASCOM NASA Communications Network
NASDA Japan National Space Development Agency
NASTRAN NASA Structural Analysis
NATO North Atlantic Treaty Organization
NAVSAT Naval Navigation Satellite
NAVSTAR Navigation System Using Time and Ranging
NBS National Bureau of Standards
ND Networks Directorate
NERVA Nuclear Engine for Rocket Vehicle Application
N_2O_4 nitrogen tetroxide
NOAA National Oceanic and Atmospheric Administration
NORAD North American Air Defense Command
NRC National Research Council
NRL Naval Research Laboratory
NTS Navigation Technology Satellite
NWC Naval Weapons Center
NWS National Weather Service
OAF Optical Alignment Facility
OAMS Orbital Attitude Maneuvering System
OAO Orbiting Astronomical Observatory
OAP Onboard Attitude Processor
OAST Office of Aeronautics and Space Technology (US)
OBP onboard processor
OCC Operations Control Center
OCCS Operations Control Center Software
ODS Orbit Determination Section
OFO Orbiting Frog Otolith
OGO Orbiting Geophysical Observatory

Saturn
Pioneer 11 (1979)
Voyager 1 (1980)
Voyager 2 (1981)

Uranus
Voyager 2 (1986)

Neptune
Voyager 2 (1989)

OMS Orbital Maneuvering System
OPF Orbiter Processing Facility
ORR Orroral
OSCAR Orbiting Satellite-Carrying Amateur Radio
OSO Orbiting Solar Observatory
OTDA Office of Tracking and Data Acquisition
OTP Office of Telecommunications Policy
OTV orbital transfer vehicle
OV orbiting vehicle
OWS orbital workshop
P&F particles and fields
PAGEOS Passive GEOS
PAM payload assist module
PCM pulse-code modulated
PDT Pacific Daylight Time
PEL precision elastic limit
PI principal investigator
PKM perigee kick motor
PM phase modulated
PMP premodulation processor
POCC Payload Operations Control Center
PSD power spectral density
psi pounds per square inch
PSK Phase-Shift Keying
PSS payload specialist station
PST Pacific Standard Time
PW Pratt & Whitney
PWM pulse width modulated
QUI Quito
RAD Rapid Access Data
R&D Research and Development
R&QA Reliability and Quality Assurance
RBM Real-Time Batch Monitor
RBV return beam vidicon
RC remote command unit
RCA Radio Corporation of America
RCS reaction control system
RF radio frequency
RF Comm Set Radio Frequency Communications Set
RFI radio frequency interference
RI Rockwell International Corporation
RMS Remote Manipulator System
RMS Radiation Meteoroid Satellite
ROM read only memory
ROS Rosman
RSS root sum square
RSU relay switching unit
RV re-entry vehicle
SAA South Atlantic Anomaly
SAC Strategic Air Command (US Air Force)
SAGE Stratospheric Aerosol and Gas Experiment
SAMD stratospheric aerosol measurement device
SAMOS Satellite and Missile Observation System
SAS Small Astronomy Satellite
SBS Satellite Business Systems
SCATHA Spacecraft Charging at High Altitudes
SCE Spacecraft Command Encoder
SCIA Spacecraft Checkout and Integration Area
SCMR surface composition mapping radiometer
SCO subcarrier oscillator
SCORE Signal Communication by Orbiting Relay Equipment
SCR selective chopper radiometer
SDM Structural Dynamics Model
SDS Satellite Data System
SECOR Sequential Collation of Range
SEM space environment monitor
SEP European Propulsion Company (France)
SERT Space Electric Rocket Test
SES space environment simulator
SETI Search for Extra-Terrestrial Intelligence
SHF super high frequency
SIO special input/output
SIR shuttle imaging radar
SIRS satellite infrared spectrometer
SIS Satellite Interceptor System
SLA spacecraft-lunar module adapter
SLBM Submarine-Launched Ballistic Missile
SLV satellite launch vehicle
SM service module
SME Solar Mesophere Explorer
SMM Solar Maximum Mission
SMS stratospheric and mesopheric sounder
SMS Synchronous Meteorological Satellite
SNAP supplementary nuclear power
SOT solar optical telescope
SPACECOM Space Command (US Air Force)
SPADATS Space Detection And Tracking System
SPASUR Space Surveillance System
SPM solar proton monitor
SPMS Special Purpose Manipulator System
SPS service propulsion system
SR Sperry Rand
SRAM short-range attack missile
SRB solid rocket booster(s)
STADAN Space Tracking and Data Acquisition Network
STDN Spaceflight Tracking and Data Network
STOP Structural-Thermal-Optical Program
STS Space Transportation System
TAD Thrust Augmented Delta
TAID Thrust Augmented Improved Delta
TAN Tananarive
T&E test and evaluation
TAR test action request
TDAS Tracking and Data Acquisition System
TDPS Tracking Data Processing System
TDRE Tracking and Data Relay Experiment
TDRS Tracking and Data Relay Satellite
TDRSS Tracking and Data Relay Satellite System
TELOPS Telemetry On-Line Processing System
THIR temperature, humidity and infrared radiometer
TIP TIROS information processor
TIP Transit Improvement Program
TIROS Television Infrared Observation Satellite
TM thermal model
TNT Trinitrotoluene
TOS TIROS Operational Satellites
TOS transfer orbit stage
TOVS TIROS operational vertical sounder
TRW Thompson Ramo Wooldridge, Inc
TV television
TV test vehicle (Vanguard)
UDMH Unsymmetrical Dimethyl Hydrazine
UHF ultra high frequency
ULA Fairbanks, Alaska
USAF US Air Force
USB unified S-band
USN US Navy
UTC United Technologies Corporation
UV ultraviolet
VAB Vertical Assembly Building
VAFB Vandenberg Air Force Base
VAS visible/infrared spin-scan radiometric atmospheric sounder
VHF very high frequency
VHRR very high resolution radiometer
VHSIC very high speed integrated circuits
VL Viking landers
VOB vacuum optical bench
VTPR vertical temperature profile radiometer
WI Wallops Island, Virginia
WTR Western Test Range, Vandenberg Air Force Base
X experimental
ZOE zone of exclusion

PRINTED IN BELGIUM BY
proost
INTERNATIONAL BOOK PRODUCTION

WATERLOO HIGH SCHOOL LIBRARY
1464 INDUSTRY RD.
ATWATER, OHIO 44201

The encyl. of U.S. spacecraft
Yenne, Bill
16765
629.4 Yen